Further Perspectives
in Organic Chemistry

The Ciba Foundation for the promotion of international cooperation in medical and chemical research is a scientific and educational charity established by CIBA Limited – now CIBA-GEIGY Limited – of Basle. The Foundation operates independently in London under English trust law.

Ciba Foundation Symposia are published in collaboration with Elsevier Scientific Publishing Company, Excerpta Medica, North-Holland Publishing Company, in Amsterdam.

Elsevier/Excerpta Medica/North-Holland, P.O. Box 211, Amsterdam

Symposium on Further Perspectives in Organic Chemistry, London, 1977.

Further Perspectives in Organic Chemistry

Ciba Foundation Symposium 53 (new series)
To commemorate Sir Robert Robinson and his research

QD241
S94
1978

1978

Elsevier · Excerpta Medica · North-Holland
Amsterdam · Oxford · New York

©*Copyright 1978 Ciba Foundation*

All rights reserved. No part of this publication may be reproduced or transmitted in any form or by any means, electronic or mechanical, including photocopying and recording, or by any information storage and retrieval system, without permission in writing from the publishers.

ISBN 0-444-90001-2

Published in January 1978 by Elsevier/Excerpta Medica/North-Holland, P. O. Box 211, Amsterdam and Elsevier North-Holland Inc., 52 Vanderbilt Avenue, New York, N.Y. 10017.

Suggested series entry for library catalogues: Ciba Foundation Symposia.
Suggested publisher's entry for library catalogues: Elsevier/Excerpta Medica/North-Holland

Ciba Foundation Symposium 53 (new series)

220 pages, 185 figures, 10 tables

Library of Congress Cataloging in Publication Data

Main entry under title:

Further perspectives in organic chemistry.

    Bibliography p.
    Includes index.
    1. Chemistry, Organic—Congresses. 2. Robinson, Robert, Sir, 1886–1975.
QD241.F87   547   13720
ISBN 0-444-90001-2

Printed in the Netherlands by Mouton, The Hague

Contents

G. W. KENNER Chairman's opening remarks 1

LORD TODD Introduction 3

A. J. BIRCH Biosynthesis in theory and practice: structure determinations 5
  Discussion 20

A. R. BATTERSBY Ideas and experiments in biosynthesis 25
  Discussion 42

SIR DEREK BARTON and S. V. LEY Design of a specific oxidant for phenols 53
  Discussion 60

R. RAMAGE Synthesis of sesquiterpenoids of biogenetic importance 67
  Discussion 79

J. BALDWIN Rules for ring closure 85
  Discussion 92

General Discussion I: Peptide synthesis 101

M. J. S. DEWAR Some recent developments in quantum organic chemistry 107
  Discussion 122

G. E. EVANS, M. J. GARDON, D. A. GRIFFIN, F. J. LEEPER and J. STAUNTON Biomimetic syntheses of phenols from polyketones 131
  Discussion 144

J. M. BROWN Selective homogeneous and heterogeneous catalysis   149
  Discussion   171

R. BRESLOW Studies on enzyme models and on the enzyme carboxypeptidase A   175
  Discussion   186

General Discussion II: Synthesis of antileukaemic lignans   191
  Organometallic synthesis   194
  Anti-aromatic compounds   201

LORD TODD Summing up   203

Index of contributors   205

Subject index   207

Participants

Symposium on Further Perspectives in Organic Chemistry (to commemorate Sir Robert Robinson and his research) *held at the Ciba Foundation on the 16th and 17th February, 1977*

Chairman: G. W. KENNER Department of Chemistry, Liverpool University, PO Box 147, Liverpool L69 3BX, UK

R. BAKER Department of Chemistry, The University, Southampton SO9 5NH, UK

J. BALDWIN Chemistry Department, Massachusetts Institute of Technology, 77 Massachusetts Avenue, Cambridge, Massachusetts 02139, USA

SIR DEREK BARTON* Department of Chemistry, Imperial College of Science and Technology, Imperial Institute Road, London SW7 2AY, UK

A. R. BATTERSBY University Chemical Laboratory, Lensfield Road, Cambridge CB2 1EW, UK

A. J. BIRCH The Research School of Chemistry, The Australian National University, PO Box 4, Canberra, ACT, Australia

R. BRESLOW Department of Chemistry, Columbia University, Morningside Heights, New York, New York 10027, USA

J. M. BROWN Department of Organic Chemistry, Dyson Perrins Laboratory, South Parks Road, Oxford OX1 3QY, UK

SIR ERNST CHAIN Department of Mechanical Engineering, Imperial College of Science and Technology, Exhibition Road, London SW7 2BX, UK

SIR JOHN CORNFORTH School of Molecular Sciences, University of Sussex, Falmer, Brighton, Sussex BN1 9QJ, UK

* *Present address:* Institut de Chimie de Substances Naturelles, Centre Nationale de la Recherche Scientifique, 91190 Gif-sur-Yvette, France

LADY CORNFORTH 'Saxon Down', Cuilfail, Lewes, East Sussex BN7 2BE, UK

M.J.S. DEWAR Department of Chemistry, University of Texas, Austin, Texas 78712, USA

A. ESCHENMOSER Laboratorium für organische Chemie der ETH-Z, Universitätstrasse 16, CH-8006 Zurich, Switzerland

B.T. GOLDING Department of Molecular Sciences, Warwick University, Coventry, Warwickshire CV4 7AL, UK

SIR EWART JONES Dyson Perrins Laboratory, South Parks Road, Oxford OX1 3QY, UK

G.W. KIRBY Department of Chemistry, Glasgow University, Glasgow G12 8QQ, UK

F. MCCAPRA School of Molecular Sciences, University of Sussex, Falmer, Brighton, Sussex BN1 9QJ, UK

V. PRELOG Laboratorium für organische Chemie der ETH-Z, Universitätstrasse 16, CH-8006 Zurich, Switzerland

R. RAMAGE* Department of Organic Chemistry, Liverpool University, PO Box 147, Liverpool L69 3BX, UK

R. A. RAPHAEL University Chemical Laboratory, Lensfield Road, Cambridge CB2 1EW, UK

C.W. REES Department of Organic Chemistry, Liverpool University, PO Box 147, Liverpool L69 3BX, UK

F. SONDHEIMER Department of Chemistry, University College London, Gower Street, London WC1E 6BT, UK

J. STAUNTON University Chemical Laboratory, Lensfield Road, Cambridge CB2 1EW, UK

LORD TODD Master's Lodge, Christ's College, Cambridge CB2 3BU, UK

R.B. WOODWARD Department of Chemistry, Harvard University, 12 Oxford Street, Cambridge, Massachusetts 02138, USA

Editors: RUTH PORTER *(Organizer)* and DAVID W. FITZSIMONS

---

* *Present address:* Department of Chemistry, University of Manchester Institute of Science and Technology, P.O. Box 88, Sackville Street, Manchester M60 1QD, UK

Chairman's opening remarks

G. W. KENNER
Department of Organic Chemistry, Liverpool University

Although this symposium commemorates Sir Robert Robinson and his researches, I shall not give a biographical introduction in view of the splendid memoir by Lord Todd and Sir John Cornforth (1976) and Sir Robert's own *Memoirs of a Minor Prophet*. Nevertheless I cannot resist mentioning his connection with Liverpool, where he occupied the newly created Heath Harrison Chair of Organic Chemistry from 1915 to 1920. It was in this period that he achieved his synthesis of tropinone. Our institute is named after him, and the photograph, which he gave us, is a constant, salutary reminder of the standard expected from the Robert Robinson Laboratories.

The title of this symposium—Further Perspectives in Organic Chemistry—was a stroke of inspiration by Professor Battersby, alluding to the famous *Perspectives in Organic Chemistry,* which Lord Todd edited (1956) in commemoration of Sir Robert's 70th birthday. The money raised by the sale of that book was used to establish the Robert Robinson Lectureship of the Chemical Society. I am delighted that six of the contributors to the earlier printed symposium are at this meeting and also that we have a 'full house' of Robert Robinson Lecturers.

The stimulus to hold this symposium arose partly from my concern about the fragmentation of organic chemistry through the formation of specialized groups, dealing with nucleotides, carbohydrates, reaction mechanisms and so forth. In themselves these groups are most useful to the specialists concerned, but this development seems to me to be contrary to Sir Robert's outlook: he was interested in so many things; for instance, Todd & Cornforth's biographical memoir mentions 11 different areas of work. His contributions to the development and application of electronic theory were comparable in importance to his contributions to synthesis. By splitting into groups, we may be in danger of missing the unity of organic chemistry. I hope that this symposium may counter that danger to some extent.

The topics in this symposium do not represent accurately in emphasis Sir Robert's own work. In particular we shall lay greater emphasis on biosynthesis and less on synthesis; that is not inappropriate, because his speculations about biosynthesis set out in the Weizmann Lectures (delivered 1953, published 1955) were so fruitful and stimulating. Structural analysis by classical chemical degradation is, of course, missing, and there must be some regret over the loss of this source of new reactions, despite the enormous advantages conferred by spectroscopic and crystallographic methods. I believe that the papers to be presented reflect the interests which he would have had in the current era of organic chemistry.

References

ROBINSON, SIR ROBERT (1955) *The Structural Relations of Natural Products*, Oxford University Press, London

ROBINSON, SIR ROBERT (1976) *Memoirs of a Minor Prophet*, vol. 1, Elsevier, Amsterdam, Oxford and New York

TODD, LORD & CORNFORTH, J. W. (1976) Robert Robinson, in *Biographical Memoirs of Fellows of The Royal Society*, vol. 22, pp. 415–527, The Royal Society, London

Introduction

LORD TODD
Christ's College, Cambridge

This symposium had its origins in some discussions Professor Kenner had with me and subsequently with the late Sir David Martin. Professor Kenner reminded me that it is now 21 years since I produced *Perspectives in Organic Chemistry* to mark Sir Robert Robinson's 70th birthday. *Perspectives* contained a series of essays by pupils and friends of Sir Robert on the current position in a variety of fields, and Professor Kenner suggested that now might be a proper time to look at some of these fields afresh by bringing together not only Sir Robert's associates but also younger workers for informal discussions. The outcome of these early talks and especially of the hard work and enthusiasm of Professor Kenner is to be seen in the programme of this symposium. Our membership covers a wide span of activities and ages; we range from specimens like Professor Prelog and myself who reach back almost to the medieval period of organic chemistry down to young men to whom Sir Robert was known only by name. Only a few of us — Professor Birch, Professor Dewar, Sir John Cornforth and myself — actually worked in the laboratory with Sir Robert, but everyone here has been influenced in one way or another by Sir Robert's work. By coming together for this symposium we seek to stimulate further progress in the various fields of organic chemistry we represent. Published under the title *Further Perspectives in Organic Chemistry* we hope that it may provide an appropriate successor to *Perspectives* and a worthy tribute to the memory of a great chemist.

Topics to be discussed in our symposium include synthetic methods, biomimetic studies and biogenesis, biosynthesis and theoretical organic chemistry — a wide range of subjects in all of which Sir Robert was interested. It would be difficult to find a single basis for the variety of individual research topics he pursued, but one of the things about his work which he himself often stressed to me and which many who did not know him personally seldom realize is that a passionate interest in colour in organic compounds, stimulated perhaps by his early work on brazilin, played a large part in his selection of research topics throughout his career.

The structural elucidation of natural products and synthesis as a confirmation of structure—these were the essence of the work of chemists of Sir Robert's generation, and his theoretical interests stemmed from them. And yet it has always seemed to me that he was more interested in the development of new methods than in the aesthetics of synthesis on the grand scale; in this he differed from Professor Woodward. He never enjoyed the hard slog of a lengthy synthesis; he preferred to do a few experiments on one reaction and move on to another. That is why his name is far more frequently associated with synthetic methods than with complete syntheses of complex natural products.

To me one of the striking things about his work on structural determination—and one which I learnt from him—is his use of analogue synthesis as a tool to be used as an adjunct and in part a replacement for classical degradative studies. A good example of this is to be found in his anthocyanin and anthoxanthin studies. True, analogue synthesis could be a dangerous weapon in the hands of the unwary, but despite his impetuosity nobody could seriously apply the adjective 'unwary' to Sir Robert.

It is, I think, difficult for the young chemist of today to appreciate fully Sir Robert's achievements; we always underestimate the achievements of our predecessors. I am sure that most students when they read, as I did, about Körner's orientation of substituents in the benzene ring must feel, mistakenly, that it was easy to make one's name in his day. Sir Robert was an organic chemist in the classical mould and structures were determined by him using the simplest equipment and without any of the physical aids we take for granted today. There was no n.m.r. and indeed scarcely any spectroscopy at all, no chromatography and little or no X-ray analysis during almost his entire career. Against that background his achievements were prodigious and his influence on all of us here perhaps greater than we realize.

Biosynthesis in theory and practice: structure determinations

A. J. BIRCH

Research School of Chemistry, Australian National University, Canberra

Abstract Classical approaches to structure determinations of natural molecules can be assisted by biosynthetic considerations. Application of these needs some structural evidence, based on degradations and spectra, as to biogenetic type. This evidence can be effectively supplemented by incorporations of precursors labelled with ^{14}C, ^{13}C or ^{15}N, as appropriate, with examinations of the distributions of label in the product. Several examples of both classical and incorporation techniques are reviewed.

Some uses of structure–biosynthetic relations have recently been summarized (Birch 1976). One use is to suggest possible structures of natural molecules, on the basis of incomplete evidence from other sources. Historically, the isoprene rule has been one of the most critical factors in determining terpene structures. I shall consider some individual cases in relation to various types of available evidence and ways of thinking. These examples are from my own work, partly because authors of papers often neglect to record the mental steps which lead them to a final result.

A CONGENER RELATION: ACORIC ACID

Congeners may be rationally derivable in many instances from common precursors, although the usefulness of this in structure determination depends on how far back in biosynthetic sequences the routes diverge.

A simple example is acoric acid (1) (Birch *et al.* 1964a). Chemical degradations suggested (1) or (2) as alternative structures but could not distinguish between them. As acoric acid occurs with acorone (3) in *Acorus calamus* L., the structure (2) is almost certainly correct, derivable from (3) by oxidation. This assumption was confirmed initially by partial synthesis from (3) and eventually by total synthesis (Birch *et al.* 1972a). The steric configuration suggested by (3) is also correct.

RELATIONS BASED ON CARBONIUM ION MECHANISMS

Neocembrene-A

This trail substance of the termite *Nasutitermes* (Birch *et al.* 1972b) is an oil, which cannot be examined by X-ray crystallography, and 2 mg was obtained from 20 kg of termites. It was accordingly examined chiefly by spectroscopic methods (mass and n.m.r. spectroscopy), its only chemical reactions being hydrogenation and degradation by microozonolysis combined with gas chromatography which showed that the molecule, $C_{20}H_{32}$, had four double bonds and gave rise to 1 mol of formaldehyde and 2 mol of laevulinaldehyde $CH_3 \cdot CO \cdot CH_2 \cdot CH_2 \cdot CHO$. Spectra showed it to contain $-CMe=CH_2$, $-CH=CMe-$ and to be monocyclic with four double bonds. The C_{20} formula, the degree of unsaturation found and the spectroscopic evidence suggested that it has an isoprenoid skeleton. The biogenetic key to the structure seemed to be the $MeC=CH_2$ group. Limonene (5) results from the cyclization of the terpene chain (4); a similar cyclization to a six-membered ring from the C_{20}-precursor geranylgeranyl pyrophosphate would not yield this group, since the resulting double bond would be trisubstituted. However, if not a six- but a 14-membered ring were formed as shown, the structure (6) would contain naturally this grouping.

The suggestion of such a ring was supported by the fact that the n.m.r. shifts of protons in the –CH=CMe– groups are slightly different to those in similar six-membered-ring analogues. Formula (6) having been suggested, it was easy to show that it is correct because key degradation reactions became immediately apparent. All these operations consumed 1.4 mg of the 2 mg of material available.

Pleuromutilin

To apply biogenetic ideas we must have some clues about biosynthetic type, even as little as the isoprenoid formula $[C_5H_8]_n$ above. In other cases usually some degradation evidence is available. We decided in several cases as an adjunct to, or a substitute for, degradation to use tracer incorporation from known simple biogenetic precursors as evidence. This would have the dual purpose of showing what units are present and possibly, by specific labelling and degradation, how the units are related to each other. The first example I shall discuss is pleuromutilin.

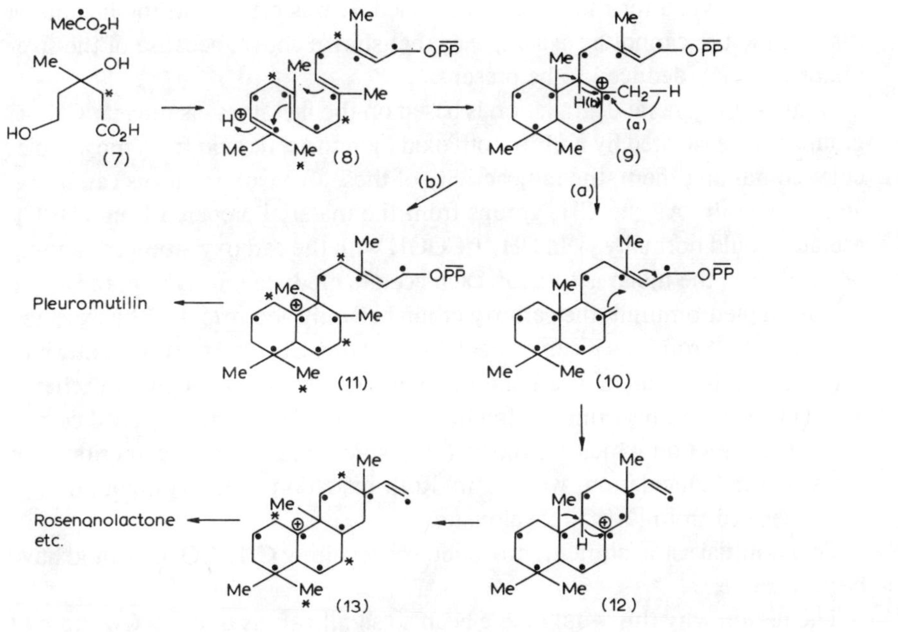

SCHEME 1. Biosynthesis of pleuromutilin from [1-^{14}C]acetate (·) and [2-^{14}C]mevalonate (*) (7).

This mould product is a glycollate of an alcohol $C_{20}H_{22}O_3$ (Birch et al. 1963, 1966), the formula alone suggesting a terpenoid origin. Infra-red spectra, supported by n.m.r. spectra when these later became available, strongly suggested the presence of a $CH=CH_2$ group. This was initially the only structural evidence available besides i.r. evidence for an OH group and a cyclopentanone ring. The group $MeC-CH=CH_2$ was known to be characteristic of several diterpenes generated by carbonium ion cyclizations of terpene chains as originally suggested by Wenkert and by Stork.

Using [1-^{14}C]acetate or [2-^{14}C]mevalonate (7) as precursors, we found the skeletal labelling and biogenetic processes outlined in Scheme 1.

We had not long previously confirmed for the first time the predicted tricyclic and tetracyclic diterpene route (Birch et al. 1959) using incorporations of labelled acetate and mevalonate into gibberellic acid and rosenonolactone, finding the expected labelling of $MeC-CH=CH_2$.

The group $CH=^{14}CH_2$ in material derived from $CH_3\cdot^{14}COOH$ should on this picture contain one eighth of the radioactivity of the molecule. This was found to be true for pleuromutilin by ozonolysis to formaldehyde. Also, the terpene precursor [2-^{14}C]mevalonate was well incorporated, although we did not degrade the product. Therefore, we were sure that it is a terpene and that biogenetic ideas valid for that area can be used. It was clear from the beginning that the picture cannot be as simple as that shown above, because of the five-membered ring deduced to be present.

A favourite general degradation is based on the fact, that, as nuclear CH_3-C groups can be isolated by Kuhn–Roth oxidation to acetic acid from most molecules containing them, the radioactivity of these two carbon atoms can be examined readily. All the CH_3 groups from the material prepared from [1-^{14}C]-acetate should normally yield $CH_3\cdot^{14}COOH$ with the carboxy group containing one eighth of the molar activity in each acetate produced. In fact, in the acetic acid from pleuromutilin the carboxy group had only about 75% of the expected radioactivity. From this it can be deduced that one CH_3 has moved during biosynthesis as the result of a carbonium ion rearrangement, as shown in Scheme 1 for (11) and (13), resulting in attachment of this CH_3 to an unlabelled carbon atom. One reaction which we did not do, which would have been critical for the structural elucidation, was Kuhn–Roth oxidation of the compound originally derived from [2-^{14}C]mevalonate.

To our initial astonishment, no doubt, the resulting $CH_3\cdot COOH$ would have been inactive.

The reason why this would have been so significant is that the C-2 atom of mevalonate should label only one terminal methyl group of a terpene chain or one only of the *gem*-dimethyl groups in a skeleton derived from it, e. g. as in

(11) and (13). The experiment that we did not do would have shown that no such labelled methyl group is present and, therefore, that the methyl group must have disappeared in a biosynthetic step. This could have been as the result of oxidation, but it was known by then from the chemistry that no primary alcohol or ester was present in the molecule.

The biosynthesis in Scheme 2 was demonstrated by later work. Although lacking this vital clue, we were considering formula (14) amongst several others on both chemical and biosynthetic grounds when Arigoni (1962) published it, having ascertained it on the basis of extensive chemical degradations.

We were able to limit the carbonyl group to one of four positions related to the original biosynthetic chain by examining the label of the carbon atom of C=O. It was converted into MeC–OH by methylmagnesium iodide, followed by Kuhn–Roth oxidation to $CH_3 \cdot COOH$. This acetic acid was radioactive when derived from material biosynthesized from [2-^{14}C]mevalonate material but not when derived from [1-^{14}C]acetate material. The carbon atom, therefore, arises from the C-2 of mevalonate, and is therefore limited to at most four and probably three positions related to the initial chain. The formula (14) agrees with this conclusion. This approach to limiting situations of groupings in terpenoid and steroid molecules, particularly in mould metabolites where tracer incorporation is easy, could be used more widely, especially with the advent of ^{13}C n.m.r. spectroscopy.

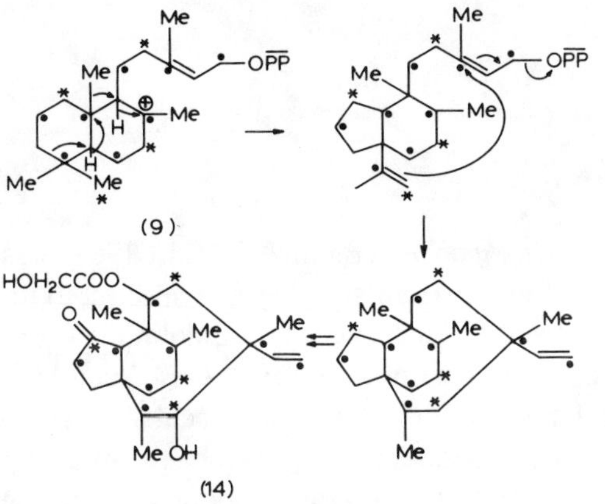

SCHEME 2

RELATIONS BASED ON SPECTROSCOPIC AND DEGRADATION EVIDENCE COMBINED WITH TRACER INCORPORATIONS

Echinulin

When we began work on it, this mould product had been assigned the partial formula (15) (Quilico *et al.* 1958) on the basis of chemical evidence. Biosynthetic considerations led us to the full structure (16) (Birch *et al.* 1961a; Birch & Farrar 1963). This was based on probable origin from tryptophan and alanine and three introduced C_5-terpene units, and needed revision of the original empirical formula by addition of CH_2. Incorporation of ^{14}C confirmed the expected origin. Formula (16) was independently shown to be correct on chemical grounds (Casnati *et al.* 1962). When this work was done, mass spectra were not available and only a few n.m.r. spectra, of key compounds only, were available — with difficulty.

Brevianamide-A

This metabolite of *Penicillium brevicompactum* (Birch & Wright 1969) was available only in mg quantities and seemed, from analyses and spectra, to represent a new type of molecular skeleton. The presence of a 3-indolinone structure was established by spectroscopy and reactions. I.r. spectra also suggested a diketopiperazine ring; that could be derived from two amino acids but, as no amino acid was released by hydrolysis, subsequent changes probably involving formation of new C–C bonds may have occurred. One amino acid might be tryptophan. From the formula $C_{21}H_{23}N_3O_3$, six rings should be present, including one benzenoid, since no other double bonds could be traced, chemically or spectroscopically. The n.m.r. spectrum indicated a CMe_2 group and the mass spectrum showed a major loss of C_5H_9, which may originate as a C_5-terpene unit, as in echinulin. If this assumption is correct, the second amino acid must lack a methyl group, but should contain a ring to assist in providing the six rings needed. The prime candidate, fitting the molecular formula, is proline. Up to this stage, all the considerations about possible skeletons were pure speculations. The biosynthetic units involved were, therefore, examined by tracer incorporations which confirmed our speculations about tryptophan, proline and mevalonate. The lack of characteristic diketopiperazine CH signals in the n.m.r. spectrum and the necessity for loss of a double bond in the C_5 unit together with requirements for two more rings suggested the structure (17). Somewhat to our astonishment, this was later confirmed.

The structure we suggested explained the ^{1}H n.m.r. spectrum and the mass spectrum (for details see Birch & Wright 1970). The pertinent point here is that the spectra were too complex for us to derive a structure from them. For example, the loss of C_5H_9 in the mass spectrum involves the fission of three C–C bonds, which is readily understood from the structure of (17) but which was not suggested by it. Apart from destructive ozonolysis of the benzene ring, no meaningful chemical degradations of brevianamide-A have been done.

Later, a minor product of the mould was shown to be (16a) which is more closely related to echinulin (16) and which is also closely related to the classically expected biogenetic precursor of (16) (Birch & Russell 1972). Incorporation of cyclo(L-tryptophyl-L-proline) into brevianamide-A was later demonstrated (Birch et al. 1974).

Phomazarin

The polyketide hypothesis has proved to be extremely useful in limiting the number of possible formulae for natural products of its class and is thus, in ways, reminiscent of the use of the isoprene rule. The first practical use of this hypothesis was to correct the formula of eleutherinol from (18) to (19) (Birch & Donovan 1953).

Phomazarin had been extensively examined chemically (Kögl et al. 1945) and Kögl formulated it as (20), in which the orientation of the pyridine ring relative to the substituted benzene ring could not be defined. Our attention was first attracted by this ambiguity, since it appeared to be a polyketide molecule, and a property of the polyketide hypothesis is to assist the placing of the substitution in one ring, relative to a more distant ring, as in the example of eleutherinol (19). We first showed that the molecule arises from either eight or nine acetate units and that the COOH was originally the methyl group of acetate (Birch et al. 1961b, 1964b). In the course of this work it became necessary for chemical and spectroscopic reasons to question Kögl's formulation of the pyridine ring and to alter it to the substitution shown in the pyridone (21). Since this formula can be derived by complete head-to-tail linkage of acetate units, it was considered more probable than the alternative with the pyridone inverted. Later, after re-examination of further anomalies in Kögl's results, we had to modify the substitution of the benzene ring to that shown in (22) (Effenberger 1973).

After these alterations, we favoured the structure (23) on the basis of polyketide origin over the alternative with the inverse orientation of the end rings. A problem, however, was that interpretations of i. r. spectra based on interactions of the quinone carbonyl with the neighbouring OH and NH seemed to favour this inverted alternative. This conclusion, however, depended on some structural assumptions, notably that the pyridone ring is in the normally expected form rather than existing as the isomeric pyridol. Because of this contradiction between biosynthetic assumptions and spectroscopic interpretations, the question was further examined.

Studies of ^{13}C n.m.r. spectra suggested that phomazarin exists as the pyridol form (24) (Birch et al. 1976). To examine this point further and also to explore the relationship of the heterocyclic ring to the rest of the molecule, we used biosynthetically incorporated ^{15}N as a probe. We did this by using labelled sodium nitrate as the sole source of nitrogen in a buffer replacing the original culture solution. In its n.m.r. spectrum, the di-*O*-methylphomazarin methyl

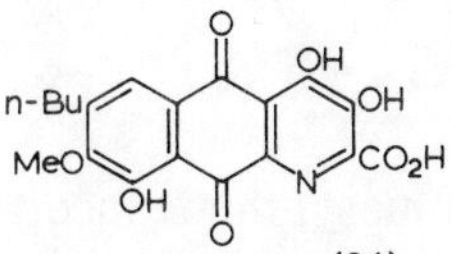

(24)

ester we obtained showed a low-field exchangeable proton at $-3.23\ \tau$ with no ^{15}N–^{1}H coupling (typically 93 Hz) and hence it was present in OH rather than NH: that is formula (24) rather than (23). This conclusion was confirmed by further n.m.r. studies.

The relative ring orientation was also studied in the same ^{15}N-enriched material. Couplings with the carbonyl carbon atoms support the pyridol structure (24) rather than the structure with the pyridol ring reversed.

The use of doubly labelled [^{13}C$_2$]acetate is now standard for examination not only of polyketide origin but of the exact dispositions of the C$_2$ units. This application depends on the fact that the enriched molecules of product are mixtures containing acetate units with either two ^{12}C or two ^{13}C-enriched carbons, in positions restricted by the origin of the chain, which can be examined by n.m.r. spectra. Studies with doubly or singly labelled acetate supplement spectroscopic studies on material with a natural ^{13}C abundance. Simpson (1975) has reviewed the methods.

If phomazarin is (24), it must arise either from two chains or, more likely, by fission of an anthraquinonoid precursor of the type (25) with considerable uncertainty about the stages of introduction of 'extra' OH and whether R is Me

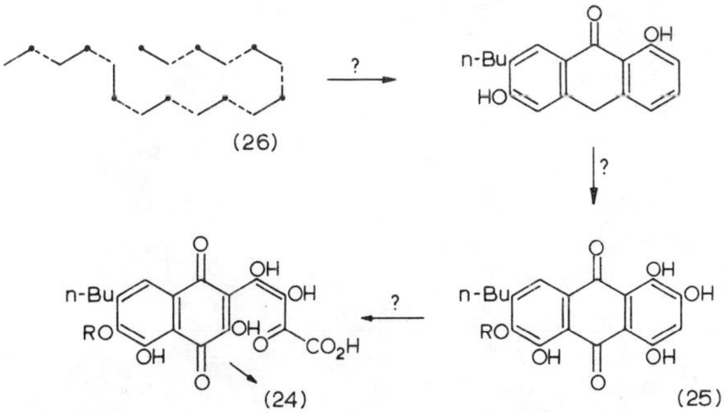

SCHEME 3

or H in a precursor (Scheme 3). This origin would also naturally explain the fact that the COOH group originates in the Me of acetate. Such experiments, including examination of acetone–malonate incorporations, support the disposition of units shown in (26).

Nystatin

The earliest, and still the most extensive, application of biosynthetic incorporations of labelled precursors in relation to the structure of a complex molecule is nystatin, an antifungal antibiotic of practical importance. The work, done in 1963–1964, was simultaneous with that on the biosynthesis of the macrolide antibiotic methymycin (Birch *et al.* 1964c). The structure of methymycin was known, and our biosynthetic work confirmed for the first time that both 'acetate' (malonyl-coenzyme A) and 'propionate' (methylmalonyl-coenzyme A) can be incorporated into a mixed polyketide chain. Nystatin, with a considerably more complex structure than methymycin, was suspected to be a polyene macrocyclic lactone. It is necessary first to note that [1- or 2-^{14}C]acetate and [1-, 2- or 3-^{14}C]propionate are efficiently incorporated into metabolites of this type with little or no 'scrambling' of the label.

When we began this work, the molecular formula was not known with certainty and was thought to be $C_{46-47}H_{73-75}NO_{16}$, and was later defined by mass spectrometry as $C_{47}H_{75}NO_{18}$ (Manwaring & Rickards 1969). Nystatin was known to contain a sugar, mycosamine, of known structure, a carboxy group, several OH groups, probably a lactone and six double bonds in separated diene and tetraene chromophores. Significant chemical evidence about the skeleton was the formation of tiglic aldehyde (27) on lead tetraacetate oxidation and of a mixture of long-chain dicarboxylic acids from fully hydrogenated nystatin, the largest one formed being (28). Nystatin itself gave only succinic acid. The effect of lead tetraacetate oxidation was then erroneously attributed to the cleavage of a *vicinal* diol; it was later established as cleavage of a structure –CH(OH)·CHMe·CH=CH– but this does not affect conclusions about the carbon skeleton.

Streptomyces noursei Dutcher synthesized ^{14}C-labelled nystatin, with good incorporations, from [1- or 2-^{14}C]acetate and from [1-, 2- or 3-^{14}C]propionate. Only minor labelling was incorporated from [Me-^{14}C]methionine and none from mevalonate. Nystatin is, therefore—at least in part—an acetate–propionate polyketide and does not contain introduced Me or terpene units.

There was initially some doubt about whether it contains three or four C–Me groups (Kuhn–Roth oxidation) but oxidation of material biosynthesized from [1-^{14}C]acetate and [^{14}C]propionate gave the active acetic acid which quantitatively agreed with four Me groups: one in the sugar (unlabelled), one from acetate and two from propionate. Since the propionate incorporations into the whole molecule indicated a total of three units, the methyl group of one of them must no longer be in that form, although its carbon atom is still present. The obvious carbon atom is that in the COOH group, and this was confirmed by decarboxylation of the [3-^{14}C]propionate-derived material to ^{14}CO$_2$. The decarboxylation needs only very mild conditions, a fact which suggests intervention of a β-keto acid or a vinylogue during the processes. Decarboxylation at much higher temperature gave a further molecule of CO$_2$, which was radioactive when the nystatin came from [1-^{14}C]acetate. This probably derives from the lactone carbonyl group. The value of the use of labelled precursors is that the results precisely indicate the specific origin of each molecule of CO$_2$ from separate sources.

The tiglic aldehyde (27), from inspection, probably contains one acetate and one propionate unit (29). This was confirmed by the fact that it contains one third of the molar activity of nystatin derived from [1-, 2- or 3-^{14}C]propionate and about 1/16th of the activity of material derived from [1- or 2-^{14}C]acetate. Nystatin therefore contains three propionate and probably 16 acetate units, from which it can be deduced that the aglycone probably contains about 41 carbon atoms. Manwaring & Rickards (1969) later established the formula to be $C_{41}H_{64}O_{15}$. The dubiety about the acetate figure relates to the fact, which we pointed out many years ago (Birch & Smith 1958), that the 'primer' unit may be incorporated to somewhat different extents. This observation was later rationalized on the basis of the first unit being acetyl-coenzyme A and the remainder malonyl-coenzyme A. In the present instance this effect is not observable. The result of high-temperature decarboxylation also roughly supported the figure of 16 acetate units but was harder to measure accurately.

The general conclusion at this stage is that the aglycone is probably totally polyketide with about 16 acetate units and with certainly three propionate units, the methyl group of one of which has been converted into COOH.

The next question was whether the labelling patterns could be used to show the order in which the biogenetic units are joined; that is, the carbon skeleton with some indications of the positions of functional groups.

Hydrogenation of nystatin and oxidation gave the dioic acid (28). Examination of propionate-labelled material showed that this acid contains two carbon atoms from the propionate COOH and one each from the propionate Me and CH_2, together with seven carbons from acetate COOH. Schmidt decarboxylation showed that one COOH arose from the carboxy group of acetate and one from that of propionate. The only reasonable deduction about labelling pattern from this, on the basis of the almost invariable head-to-tail linkage, is shown in (28). This leads in turn to the linkage of units shown in (30) with two propionate ($MeCH_2$) carbon atoms missing.

The tiglic aldehyde contains what is clearly a terminal acetate unit (i.e. it begins the biogenetic chain) and one propionate. Degradations confirm the pattern shown in (29). Is the propionate related to the propionate units of (30)? The propionate unit of (29) cannot be the complete propionate shown in (28), because if so the terminal adjacent COOH would have to arise from the acetate carboxy group which it does not. The other propionate unit of nystatin has no methyl group remaining, since this is in the form of COOH according to the experiment described before.

Vigorous alkaline hydrolysis of nystatin gave acetone and acetaldehyde, which

BIOSYNTHESIS AND STRUCTURE DETERMINATION 17

are not formed from material resulting from borohydride reduction; nor did decarboxylation occur readily after reduction. These results indicate the presence of a potential β-keto acid or a vinylogue and also that the acetaldehyde and acetone are the result of retro-aldol reactions related to the same carbonyl and, therefore, the same propionate unit. The acetone was found to contain acetate carbon atoms only (by labelling experiments) but the acetaldehyde was derived from acetate and both the C-1 and C-2 of propionate. These conclusions indicate a system such as (32) rather than (31) which explains the observed products but not their labelling. This demonstrates the significance of the use of labels, which converts an otherwise trivial result into a meaningful one. Using ^{13}C n.m.r. and mass spectrometry, we should in principle be able to distinguish between the labelling patterns in (32a) and (32b) which this result does not. The terminal carboxy group of the dioic acid (28), which is derived from C-1 of propionate, is, according to this picture, almost certainly not identical with the similar propionate carbon atom of tiglic aldehyde (27). The biogenetic skeleton can therefore probably be expanded to (33). This conclusion was confirmed by mild alkali treatment of nystatin to give (34; R=H). This was initially formulated as (34; R=OH) because the lead tetraacetate oxidation was thought to indicate a diol, and the unstable (34; R=H) was closely examined only after hydrogenation, when the OH was certainly missing (and initially thought to have been removed by hydrogenolysis). The u.v. absorption spectrum of (34) showed it to contain a pentaene-aldehyde structure, presumably also produced by retro-aldol reaction related to the same carbonyl, so the formula could now be expanded to (35) (once we have corrected, retrospectively, the erroneous OH).

(33) $\quad$ C—CO $\quad$ $\overset{\overset{C}{|}}{C}$—CO $\quad$ $\overset{\overset{C}{|}}{C}$—CO $\quad$ [C—CO]$_7$ --

(34) $\quad$ MeCHOHCHCHOH—$\overset{\overset{*Me}{|}}{\underset{R}{C}}$—CH=CHCH=CHCH$_2CH_2$[CH=CH]$_5$CHO

(35) $\quad$ MeCHCHCHOHCH[CH=CH]$_2$CH$_2$CH$_2$[CH=CH]$_4$CHOHCH$_2$CHOHCHCO$_2$H
$\qquad\qquad$ | $\qquad$ |
$\qquad\qquad$ Me $\qquad$ Me $\qquad\qquad\qquad\qquad\qquad\qquad\qquad\qquad\qquad\qquad$ CHOH
$\qquad$ O——————————————CO[C—C]$_5$—CH$_2$C—CH$_2$
$\qquad\qquad\qquad\qquad\qquad\qquad\qquad\qquad\qquad\qquad\qquad\qquad\qquad\quad$ ‖
$\qquad\qquad\qquad\qquad\qquad\qquad\qquad\qquad\qquad\qquad\qquad\qquad\qquad\quad$ O

probably mainly as [—CH$_2$CHOH—]

The lactone probably involves the OH shown, since the other one must be free because of the result of lead tetraacetate oxidation, whatever its exact mechanism. The relations shown between the COOH and carbonyl and the polyene-related CH(OH) and the carboxy group are the minimum needed by the evidence. More intervening CH(OH)·CH$_2$ units could be involved which would undergo dehydration β- to the carbonyl group and thus allow transmission of the electronic effect of the carbonyl to facilitate retro-aldol processes. The specific cleavage to (34) allows the placing of the carboxy group as in (35).

(36) MeCHCHCHOHCH[CH=CH]$_2$CH$_2$CH$_2$[CH=CH]$_4$CHCH$_2$ ... CO[CH$_2$CHOH]$_3$CH$_2$CH$_2$CHOHCHOHCH$_2$–C(CH$_2$)(OH)–CHOH–CHCO$_2$H

R = C$_6$H$_{12}$NO$_3$

After much further chemical and spectroscopic work (Chong & Rickards 1970), the structure of nystatin was defined as (36). In this formula, with one exception, the OH or equivalents are all on carbon atoms originally derived from the COOH of acetate or propionate, another type of structural feature which was predicted from the known biosynthetic origin (Birch 1967).

ENZYMIC STUDIES

The recent work showing that the antibiotic cerulenin acts as an inhibitor of fatty acid, polyketide (including propionate) and some types of steroid biosynthesis raises interesting possibilities (Omura 1976). Inhibition by this antibiotic, with minimal structural evidence, may indicate the biosynthetic type. Moreover, although it does not appear to have been examined, the use of this inhibitor might define the starter-unit of a mixed polyketide chain.

CONCLUSION

Why do this kind of investigation when X-ray crystallography is now so efficient? One answer is that X-rays are not always readily applicable because the substance is not available in appropriate form or for lack of facilities. Our approach helps to illuminate both chemistry and biosynthesis simultaneously and is also an interesting problem-solving exercise. Nowadays, with the availability of ^{13}C n.m.r. spectroscopy, ^{13}C-enriched precursors might be much more frequently used to supplement results based on natural abundance studies, in cases like mould or bacterial metabolites where appropriate incorporations can be readily obtained.

References

ARIGONI, D. (1962) Structure of a new type of terpene. *Gazz. Chim. Ital. 92*, 884–901
BIRCH, A. J. (1967) Nystatin, in *Antibiotics*, vol. 2 (Gottlieb, D. & Shaw, P. D., eds.), pp. 228–230, Springer, Berlin
BIRCH, A. J. (1976) Chance and design in biosynthesis. *Interdiscip. Sci. Rev. 1*, 215–233
BIRCH, A. J. & DONOVAN, F. W. (1953) Studies in relation to biosynthesis. III, The structure of eleutherinol. *Aust. J. Chem. 6*, 373–378
BIRCH, A. J. & FARRAR, K. R. (1963) Studies in relation to biosynthesis. Part XXXIII. Incorporation of tryptophan into echinulin. *J. Chem. Soc.*, 4277–4278
BIRCH, A. J. & RUSSELL, R. A. (1972) Studies in relation to biosynthesis XLIV. Structural elucidations of brevianamides-B, -C, -D and -F. *Tetrahedron 28*, 2999–3008
BIRCH, A. J. & SMITH, H. (1958) The biosynthesis of aromatic compounds from C_1- and C_2- units. *Chem. Soc. Spec. Publ. 12*, 1–11
BIRCH, A. J. & WRIGHT, J. J. (1969) The brevianamides: a new class of fungal alkaloid. *J. Chem. Soc. Chem. Commun.*, 644–645
BIRCH, A. J. & WRIGHT, J. J. (1970) Studies in relation to biosynthesis. XLII. The structural elucidation and some aspects of the biosynthesis of the brevianamides-A and -E. *Tetrahedron 26*, 2329–2344
BIRCH, A. J., RICKARDS, R. W., SMITH, H., HARRIS, A. & WHALLEY, W. B. (1959) Studies in relation to biosynthesis. Part XXI. Rosenonolactone and gibberellic acid. *Tetrahedron 7*, 241–251
BIRCH, A. J., BLANCE, C. E., DAVID, S. & SMITH, H. (1961a) Studies in relation to biosynthesis. Part XXIV. Some remarks on the structure of echinulin. *J. Chem. Soc.*, 3128–3131
BIRCH, A. J., FRYER, R. I., THOMSON, P. J. & SMITH, H. (1961b) Pigments of *Phoma terrestris* Hansen and their biosynthesis. *Nature (Lond.) 190*, 441–442
BIRCH, A. J., CAMERON, D. W., HOLZAPFEL, C. W. & RICKARDS, R. W. (1963) The diterpenoid nature of pleuromutilin. *Chem. Ind.*, 374–375
BIRCH, A. J., HOCHSTEIN, F. A., QUARTEY, J. A. K. & TURNBULL, J. P. (1964a) Structure and some reactions of acoric acid. *J. Chem. Soc.*, 2923–2931
BIRCH, A. J., BUTLER, D. N. & RICKARDS, R. W. (1964b) The structure of the aza-anthraquinone phomazarin. *Tetrahedron Lett. 28*, 1853–1858
BIRCH, A. J., DJERASSI, C., DUTCHER, J. D., MAJER, J., PERLMAN, D., PRIDE, E., RICKARDS, R. W. & THOMSON, P. J. (1964c) Studies in relation to biosynthesis. Part XXXV. Macrolide antibiotics. Part XII. Methymycin. *J. Chem. Soc.*, 5274–5278
BIRCH, A. J., HOLZAPFEL, C. W. & RICKARDS, R. W. (1966) The structure and some aspects of the biosynthesis of pleuromutilin. *Tetrahedron (Suppl. 8, Part II)*, 359–387
BIRCH, A. J., CORRIE, J. E. T., MACDONALD, P. L. & SUBBA RAO, G. (1972a) A total synthesis of ($\pm$)-ethyl acorate {($\pm$)-ethyl (3*RS*)-3-[(1*SR*,4*SR*)-1-isobutyryl-4-methyl-3-oxocyclohexyl]-butyrate} and ($\pm$)-epiacoric acid. An application of the generation and alkylation of a specific enolate. *J. Chem. Soc. Perkin Trans. I*, 1186–1193
BIRCH, A. J., BROWN, W. V., CORRIE, J. E. T. & MOORE, B. P. (1972b) Neocembrene-A, a termite trail pheromone. *J. Chem. Soc. Perkin Trans. I*, 2653–2658

Birch, A. J., Baldas, J. & Russell, R. A. (1974) Studies in relation to biosynthesis. Part XLVI. Incorporation of cyclo-L-tryptophyl-L-proline into brevianamide A. *J. Chem. Soc. Perkin Trans. I*, 50–52

Birch, A. J., Effenberger, R., Rickards, R. W. & Simpson, T. J. (1976) The structure of phomazarin, a polyketide aza-anthraquinone from *Pyrenochaeta terrestris* Hansen. *Tetrahedron Lett. 27*, 2371–2374

Chong, C. N. & Rickards, R. W. (1970) Macrolide antibiotic studies. XVII. Cyclic hemiketal structures in nystatin, amphortericin B, pimaricin and lucensomycin. *Tetrahedron Lett. 49*, 5053–5066

Casnati, G., Cavalleri, R., Piozzi, F. & Quilico, A. (1962) *Aspergillus glaucus* group. XVII. Echinuline. II. *Gazz. Chim. Ital. 92*, 105–128

Effenberger, R. (1973) *Applications of Organic Synthesis: Phomazarin and Prostaglandins*, Ph. D. Thesis, Australian National University, Canberra

Kögl, F., van Wessem, G. C. & Elsbach, O. I. (1945) Fungus pigments. XVI. Synthesis to explain the constitution of phomazarin. 3. *Recl. Trav. Chim. Pays-Bas 64*, 23–29

Manwaring, D. G. & Rickards, R. W. (1969) The structure of the aglycone of the macrolide antibiotic nystatin. *Tetrahedron Lett. 60*, 5319–5322

Omura, S. (1976) The antibiotic cerulenin; a novel tool for biochemistry as an inhibitor of fatty acid synthesis. *Bacteriol. Rev. 40*, 681–697

Quilico, A., Piozzi, F. & Cardoni, C. (1958) Chemical investigations of the *Aspergillus glaucus* group. XII. Echinulin. 7. *Gazz. Chim. Ital. 88*, 125–148

Simpson, T. J. (1975) Carbon-13 nuclear magnetic resonance in biosynthetic studies. *Chem. Soc. Rev. 4*, 487–522

Discussion

Birch: Some years ago, Professor Woodward, we discussed C-methylation and you said that your views about the propionate hypothesis were conditioned to some extent by the fact that it predicted terminal ethyl groups on the macrolide antibiotic chains. One antibiotic has now been found to have a terminal ethyl group, the ethyl part of which arises by C-methylation (R. W. Rickards, unpublished work). So your argument was not entirely justified although absolutely correct!

Battersby: In the Kuhn–Roth oxidation of labelled pleuromutilin, the specific activity of the acetic acid formed will depend on the yields from the various methyl groups. Some are quaternary and some are not. What is known in this case?

Birch: In those experiments there was some dilution of the C_1 unit. But in general one gets less acetate from a *gem*-dimethyl group. The method that Professor Cornforth devised (Cornforth *et al.* 1959) gives high yields of acetate from a quaternary C-Me group.

Cornforth: With cholesterol, from which one should get a theoretical yield of 4 mol of acetic acid, we got nearly 3 mol (70% yield) by refluxing it with an aqueous solution of CrO_3 for 24 h.

In explaining many biosynthetic mechanisms by carbonium ion chemistry, are we overlooking the extraordinary predilection of these compounds to form cyclopropane rings? One example in the simple terpene series is carene and another, in the biosynthesis of steroids, is presqualene. Perhaps the most curious are the cycloartenol intermediates for plant steroids. In all these cases cyclopropane rings are formed in reactions that one would not expect for a simple cation and a double bond. I wonder whether nature is indulging in some carbene chemistry without letting us know!

Birch: We were interested in the mechanisms of alkylation of aromatic rings with the isopentenyl cation and we were dubious of a direct carbonium ion reaction. We looked into the possibility of a cyclopropane intermediate (37) in the biosynthesis of echinulin. The organism that makes it produces alternative metabolites with substituents corresponding to cleavage of either the *a* or *b*

bond of such a cyclopropane ring. We fed sterically defined ^{2}H-labelled mevalonate expecting to observe loss of one deuterium atom but as this did not happen we concluded somewhat regretfully that this mechanism does not apply.

Breslow: Most experiments on carbonium ions use strong acid media in which the ions seldom deprotonate to make cyclopropanes. But carbonium ions generated in neutral media commonly make cyclopropanes. It does not seem unreasonable that the enzyme *in vivo* can catalyse the deprotonation. For instance, the reaction of *S*-adenosylmethionine with a double bond in the biosynthesis of the cyclopropane fatty acids looks like an alkylation with a deprotonation; it is hard to believe that it is either an ylide or an isolated $:CH_2$ reaction.

McCapra: In at least one case, in the α-amyrin series, one does not have to invoke an enzyme. Treatment of the dienone (38) with H_2SO_4 in Et_2O gives a cyclopropane directly; the intermediacy of a carbene is unlikely.

Woodward: I should mention that the relationship between a secondary carbonium ion and a carbene is only a matter of protonation and deprotonation.

Barton: With regard to the cycloartenol intermediates, isn't only one proton lost from the methyl group in cycloartenol?

Woodward: Yes, Arigoni has established the stereochemistry and the stoichiometry.

Baldwin: As Professor Breslow pointed out, cations can form cyclopropanes in certain conditions, such as media of low acidity in which they are relatively stable. The formation of cycloartenol in steroid biosynthesis could be explained by the fact that the cyclization enzyme, which holds within its active site the cation that is formed from the initial epoxide, has a basic site close to the methyl group of the precursor carbocation. As soon as the cation is felt at the adjacent atom, its basicity exerts its effect by removing a proton from the methyl group. For instance, the bicycloheptyl cation loses the *endo*-proton in certain conditions to give the tricyclic compound (Nickon *et al.* 1976). Another enzyme may have the basic site at a different place where deprotonation would give a different product. So it is the site at which the ion is trapped by base that may control the formation of the cyclopropyl derivatives of terpenes.

Golding: It is difficult for us to discuss the mechanism by which terpenes are formed until we know the structures of some of the enzymes involved; structural studies must obviously be an important future target (for recent progress see Croteau & Karp 1977). One may speculate that intermediates such as farnesyl pyrophosphate are bound in a highly structured environment from which water is excluded. In other enzymes, the starting conformation of farnesyl pyrophosphate will not be the same. Terpene formation is initiated by the departure of the pyrophosphate anion. A series of reactions ensues and stops when a basic group (as Professor Baldwin proposes) is suitably placed to deprotonate an intermediate. Alternatively, a conformation might be reached that allows water to enter the active site and quench the process by producing an alcohol.

Dewar: One point we should remember about enzymic reactions is that when the substrate is adsorbed on the surface of an enzyme the solvent is effectively squeezed off. It seems to me that the gas phase is probably a better model for reactions on enzymes than reactions in solution are. By this approach we can explain the high rates of reaction: in the gas phase normal ideas about acidity are turned upside down; for instance, water becomes a weak acid and toluene a strong acid. So the tendency to form carbonium ions may be much greater for a molecule adsorbed on an enzyme than one would expect from analogous reactions in solution.

The simple S_N2 reactions of halide ions with alkyl halides exemplify this. Not only do they take place in the gas phase in the way one would expect them to but the halide ion and the alkyl halide combine to form a stable pentacovalent carbon ion which we normally refer to as the S_N2 transition state. Our calculations predicted this (R. C. Bingham, F. A. Carrion & M. J. S. Dewar, unpublished work). Dougherty *et al.* (1974) found that the halide ion combines exothermically with the alkyl halide in the gas phase with large heats of reaction, to form such adducts. There is no guarantee that they have that structure, but it is likely.

The effect of the solvent in organic chemistry is probably far greater than anyone has realized in that it drastically alters reaction mechanisms. Most of the activation energy for ionic reactions in solution probably comes from solvation. Most of the simple ionic reactions of organic chemistry would take place in the gas phase without activation. This has been demonstrated by ion cyclotron resonance for several S_N2 reactions in the gas phase (Bohme *et al.* 1974), which are slow in solution. So with regard to carbonium ions one should exercise caution in drawing conclusions from results in solution, particularly with polar solvents.

Eschenmoser: Why should the gas phase be a better model for enzymic reactions than a solvent which is more protein-like than, say, water, dimethylformamide or tetrahydrofuran? The solvent is probably removed from the substrate by the enzyme but is replaced by a specific kind of solvating environment which is by no means modelled by the gas phase.

Dewar: Ions in solution are solvated by solvent molecules and, as a rule, the solvent molecule has to be removed before a reagent can approach and react with the ion. The enzyme contains neutral groups which take no part in the reaction besides the reaction centres. So, effectively, the reactants are approaching each other with no solvent between them—as in the gas phase.

Another factor is that when a neutral substrate is adsorbed on a neutral enzyme, only little energy is needed to squeeze out the solvent. In many instances reactive groups are then produced in the right place by a relay mechanism. In chymotrypsin, for example, a negative charge is transferred from a carboxylate ion *via* the imidazole ring of a histidyl group which forms part of the active site.

Woodward: That sounds to me like the Indian rope trick! The work has to be done somehow.

Dewar: In a reaction that starts from a neutral species and produces neutral species, the change in solvation energy is small but zwitterions are produced in the intermediate. When an ion reacts with a neutral molecule to give another neutral species (of similar solvation energy) and another neutral ion, the solvation energy of the transition state differs considerably because the solvent has to be pulled off the ion for it to react. As a rule, all the solvent does in an ionic reaction is interfere with the reaction. It isn't a question of solvents accelerating reactions. Some solvents have a less deleterious effect on the rate than others but the optimum conditions would dispense with solvent. Let us consider the reaction of HO⁻ with toluene: they do not react in solution but in the gas phase they react quantitatively to give water and the benzyl anion.

Ramage: A large part of a crystal of an enzyme may be water; as Professor Hodgkin described (1977), insulin crystals contain 60% water. That would have to be removed from the active site.

Dewar: It depends where the water is. I am not putting forward a theory of enzyme reactions but an interesting possibility which is suggested by the large differences between reactions in solution and those in the gas phase (for instance, the S_N2 reactions which are instantaneous in the gas phase but slow in solution). The same seems to be true for carbonyl and ester reactions, such as hydrolysis. We should keep this in mind, especially as it is difficult to see why enzymic reactions are as fast as they are. Simple model enzymes containing the reactive groups cannot equal the real enzyme in most cases.

Brown: We must consider two aspects of the generation of carbonium ions from, say, pyrophosphates at an enzymic site. The formation of the ion does not differ much from the corresponding reaction in solution, despite what Professor Dewar says. Remarkable differences are seen in the mechanism and site of trapping of that carbonium ion. Here one finds enzyme effects.

References

BOHME, D. K., MACKAY, G. I. & PAYZANT, J. D. (1974) Activation energies in nucleophilic reactions measured at 296°K *in vacuo. J. Am. Chem. Soc. 96,* 4027–4028

CORNFORTH, J. W., CORNFORTH, R. H., PELTER, A., HORNING, M. G. & POPJÁK, G. (1959). *Tetrahedron 5,* 311–339

CROTEAU, R. & KARP, F. (1977) Biosynthesis of monoterpenes: partial purification and characterisation of 1,8-cineole synthetase from *Salvia officinalis. Arch. Biochem. Biophys. 179,* 257–265

DOUGHERTY, R. C., DALTON, J. & ROBERTS, J. D. (1974). S_N2 reactions in gas-phase structure of transition-state. *Org. Mass Spectrum. 8,* 77

HODGKIN, D. C. (1977) The crystal structure of insulin (Royal Society of Medicine Gold Medal Lecture). *Proc. R. Soc. Med.,* in press

NICKON, A., LAMBERT, J. L. OLIVER, J. E., COVEY, D. F. & MORGAN, J. (1976) β-Epimerization and γ-hydrogen abstraction via homoenolate ions. *J. Am. Chem. Soc. 98,* 2593–2599

Ideas and experiments in biosynthesis

A. R. BATTERSBY

University Chemical Laboratory, Cambridge

Abstract During the past 25 years or so, there has been almost undreamed of progress in understanding the pathways by which living systems synthesize the remarkable range of substances they contain. This progress could not have been made had not isotopes of carbon, nitrogen, hydrogen and oxygen become available in quantity at a time when the intellectual climate was right for their penetrating application in biosynthetic research. It was by his generation of far-reaching ideas about biosynthesis that Sir Robert Robinson made such a major contribution to establishing this right climate. His thinking pointed the way for many studies on living systems.

Several examples will be discussed which were of particular interest to Sir Robert, such as the biosynthesis of morphine and colchicine, and another topic which is currently at a fascinating stage of development, the biosynthesis of natural porphyrins. New equipment and techniques, especially ^{13}C n.m.r. spectroscopy and high-pressure liquid chromatography, have helped in a broad study of the biochemical conversion of porphobilinogen into uroporphyrinogen-III which must be formed by some rearrangement process. It is established that a single *intramolecular* rearrangement occurs and that this step comes at the end of the assembly of four porphobilinogen units which forms the unrearranged bilane.

To speak at a meeting commemorating Sir Robert Robinson and his genius is both a great honour and a daunting task. No one could hope in a single lecture to follow the threads of his contributions through to present day research even for the one area of biosynthesis; the totality is too vast. So my aim is to select a few examples which illustrate the interaction of ideas and experiments, and I have chosen those biosynthetic problems which I remember gave him enormous pleasure as the experimental story unfolded. Initially I shall cover the ground quickly, so to speak by funicular, and later climb on foot for a detailed exploration.

Sir Robert's uncanny feeling for biosynthesis is laid out for all to see in his account of *The Structural Relations of Natural Products* (1955). His key assertion throughout these essays was that nature works by laws recognizable to the chemist; in other words, there is no magic. As he stressed repeatedly, even enzymes are unlikely to disregard stereochemistry or the mode of electronic displacements in the molecules. On this basis, he brought order to what most chemists must have seen as a jumbled collection of structures, and nowhere was he more successful than in the alkaloid field.

One of his most brilliant successes where he used the interplay between structure and biosynthesis was for the case of morphine, which at the time was of unknown structure. His conception of a relationship between morphine and the then well known benzylisoquinoline system is shown below from his own account (Robinson 1955); such ideas come to few men.

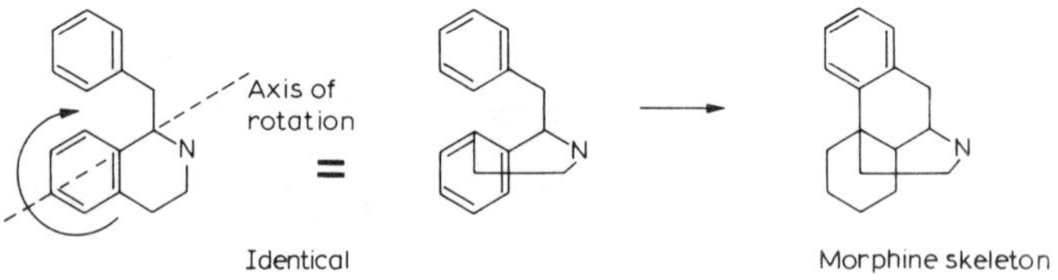

I recall being fascinated as a student at Manchester by the amazing chemistry of the morphine alkaloids but even more so by the sheer beauty of Sir Robert's idea. Was it possible, one wondered, to help solve the problem of morphine biosynthesis? Not then, but some five or six years later when carbon-14 was available at an accessible price did experimental studies become possible. It took several years to find out how to make opium poppies behave as helpful collaborators and to do the basic work with simple building blocks (Battersby & Harper 1958; Battersby et al. 1962) before we could do the key experiment, illustrated in Scheme 1 (Battersby & Binks 1960; Battersby et al. 1966). ^{14}C-labelled norlaudanosoline (1) was synthesized and was converted by the poppy plants into morphine (2) which was shown to be specifically labelled as illustrated.

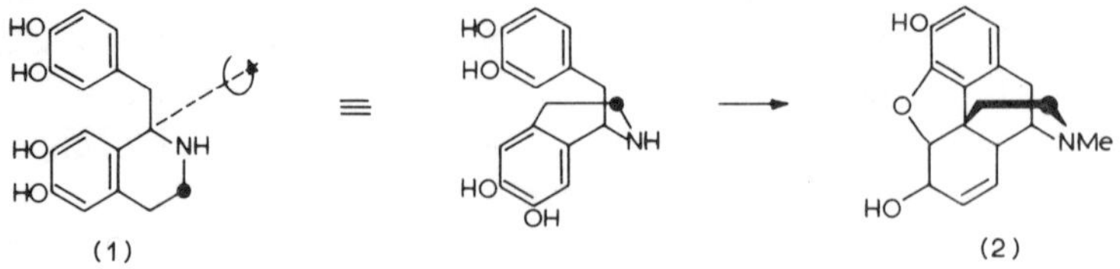

SCHEME 1.

This result established the relationship Sir Robert had proposed and it laid open the problem of morphine biosynthesis; moreover, by showing that large synthetic precursors could be successfully introduced into the plants' biosynthetic machinery, it gave the confidence to test other precursors in the opium poppy and in many other living plants.

I cannot cover here the many experiments needed by Sir Derek Barton's group and by ours to discover the detailed pathway between norlaudanosoline (1) and morphine (2) (Scheme 2) (see Battersby & Harper 1960; Battersby et al. 1963, 1965a,1967a, 1968; Barton et al. 1963, 1967). Each stage illustrated has the support of labelling studies, stereochemical correlations and isolation work; the combined evidence is overwhelming. The central part of the pathway

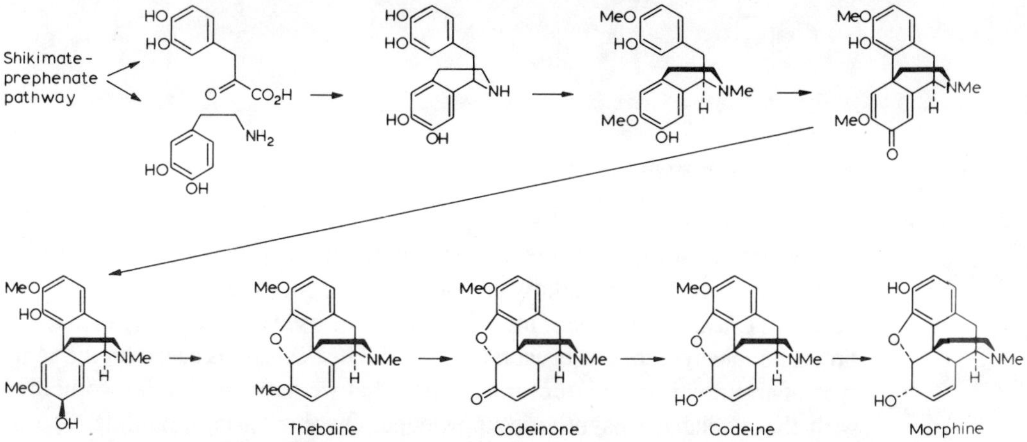

SCHEME 2. The biosynthesis of morphine.

around the dienone emerged from a fruitful collaborative effort (Barton et al. 1965; for studies with $^{14}CO_2$ see, e. g. Martin et al. 1967 and references therein) and we must never forget, in relation to the aryl–aryl coupling step, the highly important ideas of Barton & Cohen (1957) and of Erdtman & Wachtmeister (1957) on phenol oxidation.

Any organic chemist looking at the whole pathway to morphine will be filled with admiration for this synthesis. The pentacyclic system is drawn by a relatively small number of deft strokes and we see here the economy of nature's chemistry beautifully illustrated.

On this thought about economy, we should turn to the group of natural products based on the benzophenanthridine system. There had been speculations about the origin of such molecules which involved joining pieces together without any apparent relation to other alkaloids. This was contrary to Sir Robert's approach and his view (1955) was satisfying in its economy; the same idea was independently put forward by Turner & Woodward (1953). The suggestion was that the benzophenanthridine system arose by modification of an initially formed tetrahydroprotoberberine skeleton. The necessary C–N bond cleavage and C–C bond formation are indicated (without regard to mechanism at this stage) in Scheme 3.

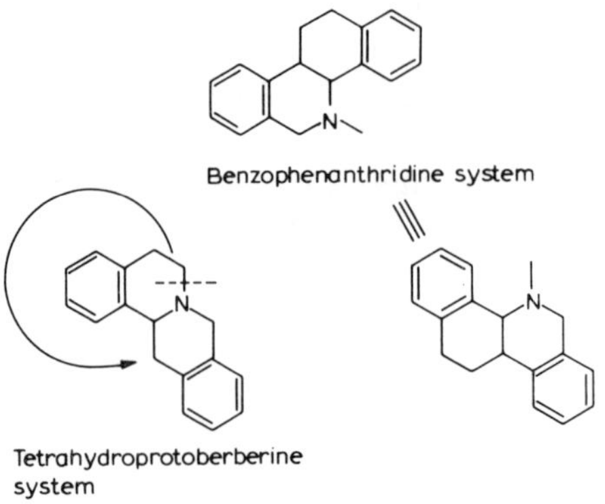

SCHEME 3.

How can we study such a relationship in the living plant? First, chelidonine (5) was adopted as a splendidly crystalline representative of this group which occurs in *Chelidonium majus;* but what should be used as precursor? That question is not always straightforward to answer. The decision is based on the same type of argument using structural relation that we have seen so far, combined with the available biosynthetic knowledge. On this basis, reticuline (3) was chosen and it was synthesized in multiply-labelled form, an approach which was extensively used in the researches on morphine. In addition, the labelled

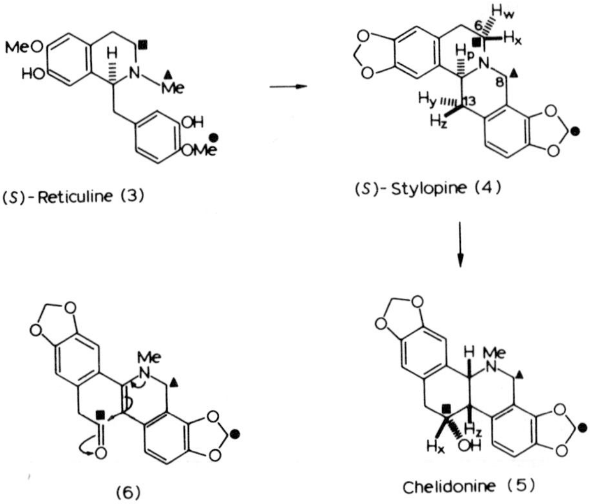

(S)-Reticuline (3)

(S)-Stylopine (4)

(6)

Chelidonine (5)

SCHEME 4.

reticuline was resolved [see (3)] to allow the rapier of stereochemistry to be added to the broad sword of structural relations.

We found by experiments on living plants (Battersby *et al.* 1965*b*, 1967*b*, 1975*a*) that the first stages of the pathway involved three ring-closures, two to form methylenedioxy groups and the third to close the so-called berberine bridge [C-9 in (4)]. (These oxidative cyclizations are by now well established for many cases by Sir Derek's group and by ours.) We isolated the (*S*)-stylopine produced and showed by suitable degradations that the labels were located at the illustrated sites. Importantly, the proportion of the total radioactivity at each position remained constant over the change (3) → (4). But it was even more satisfying for us to find that stylopine (4) was converted enzymically into chelidonine (5) and that the labelling pattern was qualitatively and quantitatively unchanged (Battersby *et al.* 1975*b*), as marked on structure (5).

So nature has generated in this way a fresh family of natural products by re-moulding the available clay, and it works with marvellous precision. Just look at the stereochemical control over the bond-breaking and bond-forming steps; by suitable isotopic substitution the centres C-6 and C-13 were made chiral and we found that the transformations were stereospecific at both sites, with H_x and H_z being retained (Battersby *et al.* 1975*c*). In addition, appropriate labelling with tritium revealed that H_p is lost and this last result (Battersby *et al.* 1975*b*) supports the idea of there being an intermediate (6) which is plausible on mechanistic grounds.

Lastly on the funicular, I want to look at colchicine, from *Colchicum autumnale,* one of six favourite problems which have given us the greatest joy (as well as a good deal of sweat and disappointments on the way). Sir Robert remarked that colchicine is *sui generis* and no structural relation with other alkaloids can

Colchicine

be recognized. With the benefit of hindsight, this opinion is not a surprising one; the secret was far too well hidden in this case. However, he did visualize the formation of simpler natural tropolones—the thujaplicins—by ring expansion in some unknown way from a benzene ring and an attached carbon, *i. e.* from an Ar–C residue (Scheme 5). The interest of this will soon become apparent.

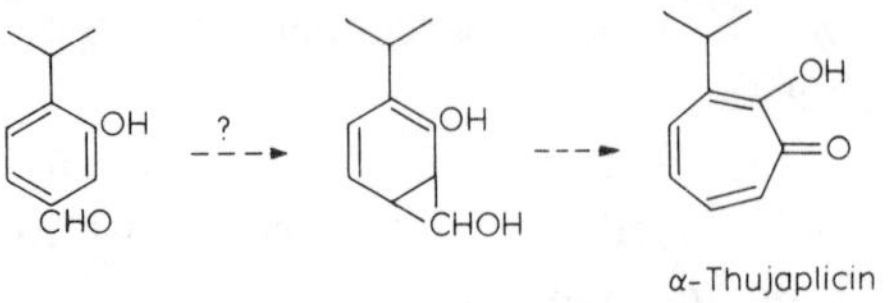

α-Thujaplicin

SCHEME 5.

The thrust during the early years of our study of colchicine (Battersby & Herbert 1964; Battersby 1967; Battersby et al. 1972a) was aimed at discovering what the building blocks were. It turned out that ring A and the attached three carbons of ring B of colchicine are derived from cinnamic acid, this in turn being produced from phenylalanine. The origin of the tropolone ring proved to be a much tougher nut to crack and it was exciting to find eventually that it is built by ring-expansion of the aromatic nucleus of tyrosine *with inclusion of the first carbon atom of the side-chain.*

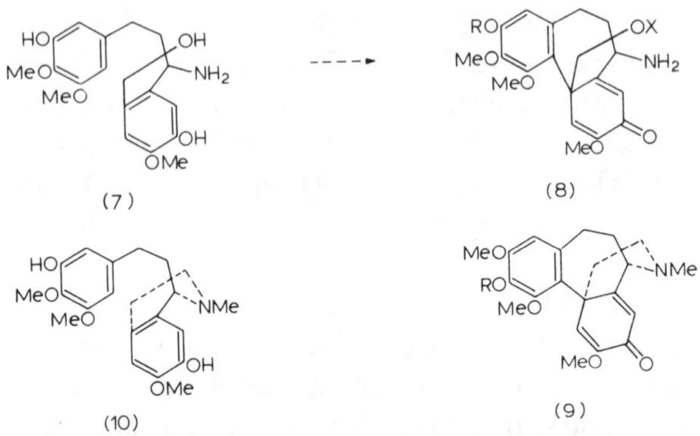

SCHEME 6.

Immediately the blackboard speculations could start and we thought that a likely precursor for colchicine might be (7) (Scheme 6) which by phenol coupling would lead to the dienone (8). We tested the diphenol (7) in *Colchicum* plants (Battersby & Herbert 1964; Battersby et al. 1972a) but the tests were abortive, correctly so as you will see. The next leap forward came from structural work on a minor alkaloid from the plant family producing colchicine which had the structure and absolute configuration (9) (Battersby et al. 1972b). Not only was the substance of great interest in being the first 1-phenethylisoquinoline alkaloid to be discovered but also because of its relation to the dienone (8) which had been envisaged as a possible precursor of colchicine on the basis of the tracer results. Now it was almost blindingly clear that everything one knew about this problem would fall perfectly into place if colchicine were biosynthesized from a 1-phenethylisoquinoline (10). And so it turned out

(Battersby 1967; Battersby et al. 1972c). The stage was thus reached where the whole beautiful pathway could not escape detection. It is shown in Scheme 7. Apart from the exact nature of the intermediate in the ring-expansion process, every step has full experimental support. For example, the key phenethylisoquinoline intermediate (10) was appropriately labelled with ^{3}H, ^{14}C, and ^{15}N to

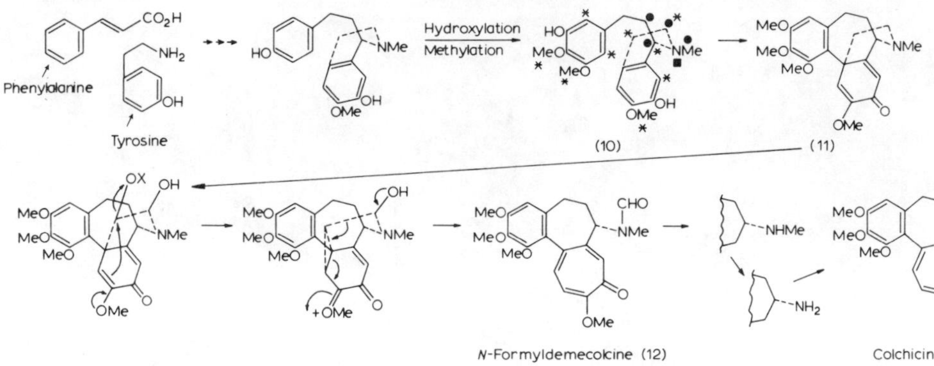

SCHEME 7. The biosynthesis of colchicine.

allow events at 12 sites in the molecule to be studied; the results were in exact agreement with Scheme 7. Also, the intermediates (10), (11), (12) and the two that follow (12) on the pathway have all been isolated from *Colchicum* plants (Battersby et al. 1971; A. R. Battersby, G. Hardy, E. McDonald, R. N. Woodhouse, D. R. Julian & R. Ramage, unpublished results).

What I find particularly fascinating about this biosynthetic pathway is that essentially the same chemistry is involved in the first half as in the biosynthesis of morphine but a phenethylisoquinoline is used rather than a benzylisoquinoline. The two pathways diverge sharply after the dienone stage, largely because of one reaction; for colchicine, the bridge is attacked next but for morphine, the carbonyl group is reduced. Here again, we see nature's economy. In this connection it is tempting to think of synthesizing the colchicine analogue of morphine and *vice versa*.

Having given a bird's eye view of three marvellous natural syntheses, I want now to look in more detail at some current researches. Sir Robert was attracted by all natural products and he was deeply interested in the major biosynthetic problem posed by the natural porphyrins. There are few more intruiging ones, as can be judged from Scheme 8. The splendid earlier work of Shemin, Granick, Neuberger, Bogorad and Rimington (for review see Battersby & McDonald 1975) had shown that the natural porphyrins are biosynthesized by way of porphobilinogen (13). The product of straightforward combination of four porpho-

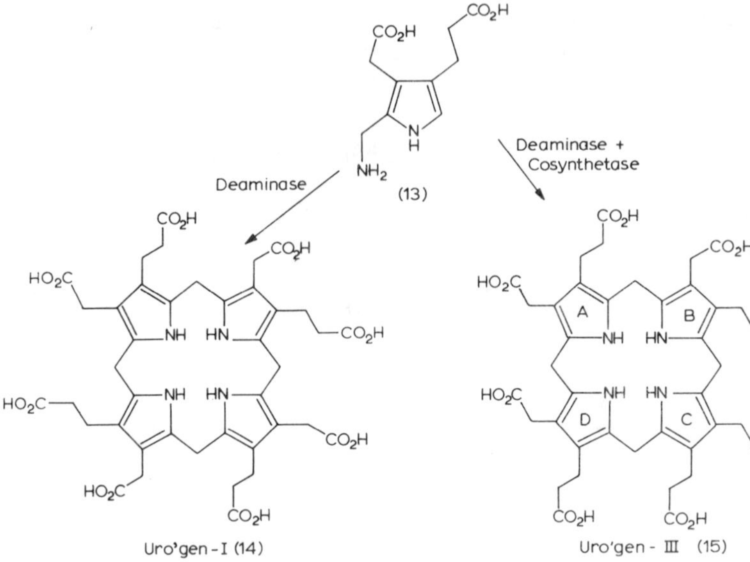

SCHEME 8.

bilinogen units head-to-tail would be uroporphyrinogen-I (uro'gen-I; 14) and this is formed when porphobilinogen is treated with the enzyme deaminase (uro'gen I synthase; EC 4.3.1.8) alone. Yet haem, the cytochromes, chlorophyll and vitamin B_{12} are all derived (see Battersby & McDonald 1975) from the unexpected isomer uro'gen-III (15). Bogorad (see Battersby & McDonald 1975) showed that this subtle change needs the cooperative action on porphobilinogen of deaminase with a second protein, cosynthetase. So living systems have evolved a highly specific way to catalyse a rearrangement process which produces the unexpected type-III system (15).

How can this problem be solved? It happens to be an unusually difficult one because, unlike the cases we have covered so far, the uro'gen-III macrocycle is built from *four identical units* and, further, the product is a highly reactive species.

The problem divides into three parts: *What* happens? *When* does it happen? *How* does it happen? As we have detailed what happens (Battersby *et al.* 1973a, 1976b), I need only briefly summarize the main points.

When one considered the many hypothetical schemes which had been proposed to account for the formation of the type-III macrocycle, it was clearly necessary to discover (*a*) how many of the four porphobilinogen units undergo rearrangement during the biosynthesis, (*b*) whether each rearrangement is *intramolecular* with respect to the porphobilinogen unit involved or whether *intermolecular* processes are operating, and (*c*) at which *site(s)* the migrating carbon atom(s) appear(s) in the final type-III macrocycle. *A priori,* the spectrum of possibilities ranges from three intramolecular rearrangements to various intermolecular processes; Sir Robert's speculations (1955) involved the former and they are shown in Scheme 9.

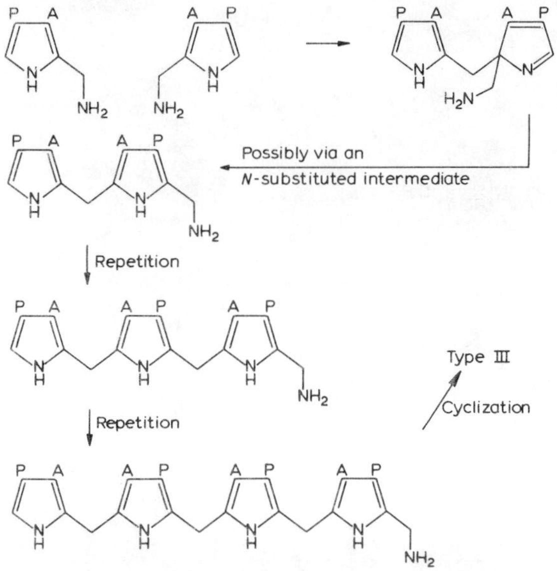

SCHEME 9.

We planned to use ^{13}C-labelling in conjunction with ^{13}C n. m. r. spectroscopy to characterize whatever bond-making and bond-breaking steps occur.

In the key experiment we synthesized [^{13}C$_2$]porphobilinogen (Scheme 10), doubly-labelled within the same molecule, and diluted it with four parts of unlabelled porphobilinogen. This dilution has the result that most uro'gen-III molecules, enzymically produced from the mixture, contain only one doubly-labelled porphobilinogen unit. So about one quarter of the uro'gen-III molecules will have ring A derived from labelled porphobilinogen, another quarter will have ring B so labelled, and similarly for rings C and D. We then showed by ^{13}C n. m. r. spectroscopy that these four major species are labelled as indicated in Scheme 10, where the illustrated labelling is *within the same molecule.*

This result (Battersby *et al.* 1973*a*, 1976*b*) is just the tip of the iceberg and I have simplified the presentation. The rest of the iceberg involved rigorous assignments of ^{13}C n. m. r. signals by synthesis (Battersby *et al.* 1973*b*), determination of the size of ^{13}C–^{13}C couplings with synthetic labelled samples (Battersby *et al.* 1976*c*), proof of integrity of the macrocycle (Battersby *et al.* 1976*a*) and so on. From Scheme 10 we can see exactly *what* has happened as the de-

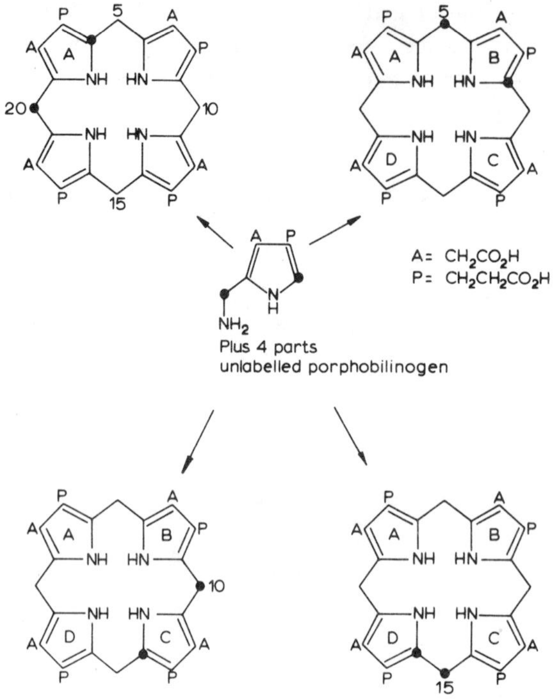

SCHEME 10.

aminase–cosynthetase system converts porphobilinogen into uro'gen-III: (*a*) the three porphobilinogen units which form ring A and its attached C-20 bridge, ring B and the C-5 bridge, and ring C with its C-10 bridge are all incorporated *intact** without rearrangement; (*b*) the porphobilinogen unit which forms ring D is built in with rearrangement which is intramolecular with respect to that porphobilinogen unit; (*c*) the rearranged carbon atom forms the bridge at C-15.

These precise requirements strictly limit the possibilities for *when* and *how* the rearrangement takes place but two sequences which fit the evidence at this stage are shown in Schemes 11 and 12. These differ in mechanism and in timing and thus we come face to face with the question *when?*

First, though, we should pause to think. We have assumed so far that uro'gen-I (see Scheme 8) is the product of uncomplicated head-to-tail joining of four porphobilinogen units and we have been mentally contrasting what

*This refers throughout to an *intact carbon skeleton;* obviously a C–N bond has been cleaved.

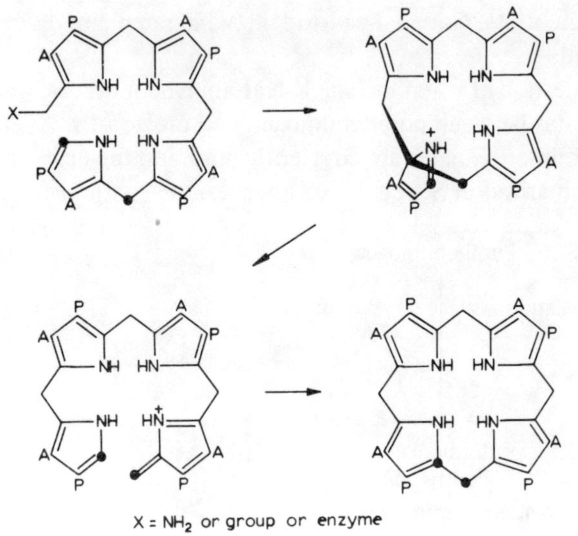

IDEAS AND EXPERIMENTS IN BIOSYNTHESIS

X = NH₂ or group or enzyme

SCHEME 11.

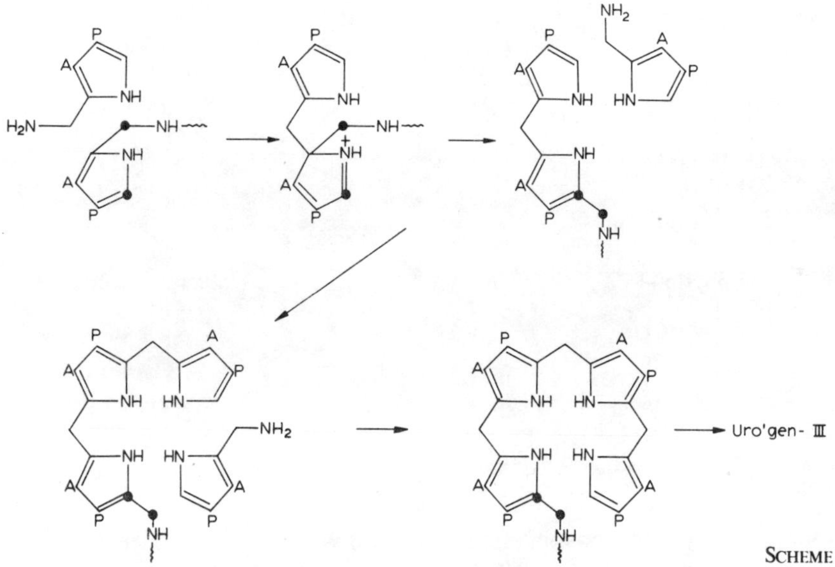

SCHEME 12.

happens for uro'gen-III with that situation. Could it be that living systems are acting unfairly by building uro'gen-I in a less-than-straightforward way? R. Hollenstein answered this question using just the same approach based on doubly ^{13}C-labelled porphobilinogen that has been outlined already. Without going into details, he found (certainly to my relief) that uro'gen-I is synthesized from four porphobilinogen units, joined head-to-tail, which remain *intact* throughout. So the ground is secure.

Returning now to the question of when the single rearrangement occurs, we must study the intermediates between porphobilinogen and uro'gen-III. As it is unknown whether these intermediates are covalently bound to the enzyme we face the situation summarized in Scheme 13. However, the synthetic di-,

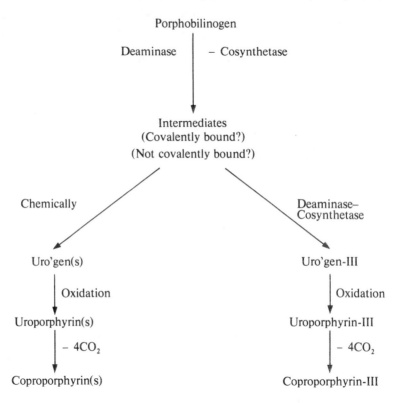

SCHEME 13.

tri- and tetra-pyrroles (possible intermediates) which *we* add in labelled form to the enzyme system will initially be free in solution and thus able to undergo competing chemical cyclization to the porphyrinogen system. The competing chemical and enzymic processes will produce mixtures of isomeric porphyrinogens. In addition, everything down the two tracks of Scheme 13 as far as and including the uro'gens is highly reactive and unstable to light, acid, oxygen and heat. You can see why we often felt like Mozart's Tamino in constant need of a magic flute.

Two essential steps had to be taken: (*a*) isolate the purified deaminase–cosynthetase enzyme system in a concentrated form so as to ensure maximum movement of material along the enzymic pathway (right-hand arrow on Scheme 13); (*b*) devise quantitative methods for the separation of isomeric porphyrins — we invested a major effort into the development of high-pressure liquid chromatography (h. p. l. c.) for this purpose. In practice, the mixture of uroporphyrins (octacarboxylic acids) is decarboxylated to the corresponding coproporphyrins (tetracarboxylic acids) before analysis of the methyl esters by h. p. l. c.; it is firmly established that the decarboxylation stage does not interconvert isomeric porphyrins. Conditions were found (Battersby *et al.* 1976*d*) for reproducible separations of all four coproporphyrin isomers (i. e. types I, II, III and IV); it is worth noting in passing that there has been considerable medical interest in these methods for studies of various porphyrias.

The rest of this story centres on several crucially important h. p. l. c. separations of isomeric coproporphyrins. The successes already described will have given confidence in the method but I want to describe one further example to strengthen this feeling and to emphasize its amazing fractionating power when we use the technique at its limit. M. Thompson mixed the protiomethyl ester of uroporphyrin-II with the corresponding deuteriomethyl ester and then fractionated the mixture in our best h. p. l. c. conditions. Separation was soon evident and was essentially complete after 12 recycles; mass spectrometry then confirmed the clean separation of the deuterio from the protio species. That is the power which is being applied in all that follows.

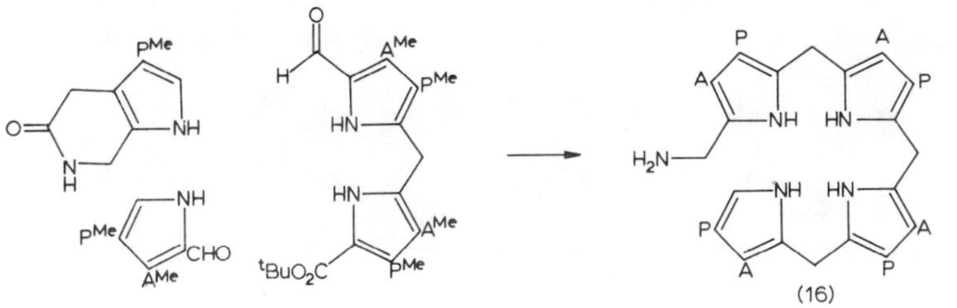

SCHEME 14.

To simplify the presentation, I shall concentrate on one set of experiments. The strategy was to synthesize the aminomethylbilane (16) as shown in Scheme 14; this bilane would be formed by straightforward *head-to-tail* reaction of four porphobilinogen units. Initially the aminomethyl group is protected as the lactam and is released in a final hydrolytic step. At pH 7.2 (which matches what will be used for the enzymic runs later) this bilane (16) ring closed chemically by warming at 55°C and the product was shown, by the methods outlined above, to be >95% pure uro'gen-I (14) (Scheme 15).

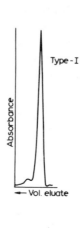

SCHEME 15.

This experiment gives two messages. The first is that the synthetic bilane has the sequential head-to-tail structure which should have been generated by the synthetic route. The second is very important and tells us that no significant rearrangement occurs at pH 7.2 as the aminomethylbilane (16) cyclizes *chemically* to the uro'gen macrocycle.

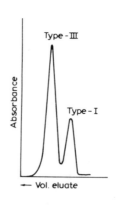

SCHEME 16.

The crucial experiment can now be run in which the aminomethylbilane (16) is treated with the purified deaminase–cosynthetase enzyme system at pH 7.2. The product was worked up and analysed as previously and the striking result is shown in Scheme 16; the major product is now type III. A little more evidence is still needed to make this statement secure; the h. p. l. c. conditions we used do not separate type-III from type-IV coproporphyrin esters. Therefore, the left-hand peak was isolated preparatively and re-run in conditions which will distinguish III from IV; the outcome was that a negligible quantity of coproporphyrin-IV (0–<5%) was present.

This experiment demonstrates a major enzymic conversion of the sequential head-to-tail bilane (16) into uro'gen-III (15) and the proportion 70:30 of type-III: type-I shows an enzymic rate enhancement relative to the competitive chemical ring-closure (Battersby et al. 1977a).

This work with the unrearranged bilane (16) has been based so far on unlabelled material. By a suitable labelling experiment we ought to be able to add further strength and penetration to the study; e.g. evidence should be provided against possible breakdown of the bilane into fragments which are incorporated into uro'gen-III. We used specific labelling with ^{13}C as illustrated in Scheme 16 and showed by n.m.r. spectroscopy that the uro'gen-III produced enzymically from it was labelled at C-15 (Battersby et al. 1977a). This not only eliminates significant breakdown and scatter of the label but also registers the bilane with respect to uro'gen-III and the corresponding rings are lettered (Scheme 16).

This brief outline of the key experiments summarizes our evidence for enzymic conversion of the unrearranged bilane (16) into the type-III macrocycle (15). The conclusion from this part of the research is that the single intramolecular rearrangement affecting ring D of uro'gen-III comes at the end of the assembly process. In this connection, it is important to record the results from Professor Müller's group (Dauner et al. 1976) who found that deaminase–cosynthetase from *P. shermanii* converted unlabelled bilane (16) (prepared in a different way from ours) into a mixture of 14–18% uro'gen-III and 82–86% uro'gen-I. Their conversions into the type-III system are much lower, it is true, probably because of enzyme concentration but nevertheless their results reinforce ours.

We have also put much effort over many years into the enzymic incorporation of aminomethylpyrromethanes into uro'gen-III. The results (Battersby et al. 1977c, d) add much further strength to those I have outlined for the bilane (cf. Frydman & Frydman 1975; Scott et al. 1976).

So we can build on knowledge of *what* and *when* in planning the attack on *how* the rearrangement takes place and the future holds great interest. For example, two further doubly ^{13}C-labelled forms of the unrearranged bilane (16) are being synthesized (by M. J. Meegan and C. Fookes) to gain additional evidence for intact incorporation of the bilane molecule and to provide the base for the next advance.

All the evidence I have presented points to Scheme 11. Evidence for the possible spiro intermediate would give enormous support and we are seeking it. Still further ahead is (*a*) an understanding of the interaction between deaminase and cosynthetase and (*b*) the crystallization and X-ray structure determination of deaminase.

All this lies in the future, but for the researches aimed at a precise chemical understanding of the biosynthesis of natural porphyrins, the long road is nearing an end.

ACKNOWLEDGEMENTS

The examples I have chosen were from researches starting in 1953. The knowledge gained in these cases and for several others, such as the indole alkaloids and vitamin B_{12}, depended on contributions from a succession of first-class young scientists. Without them, today's story could not have been told. Their names are given in the references. I am extremely grateful to them all and especially to my two senior partners, Dr. Edward McDonald and Dr. James Staunton who have been towers of strength throughout.

References

BARTON, D. H. R. & COHEN, T. (1957) in *Festschrift für Arthur Stoll*, p. 117, Birkhauser, Basle
BARTON, D. H. R., KIRBY, G. W., STEGLICH, W. & THOMAS, G. M. (1963) *Proc. Chem. Soc.*, 203
BARTON, D. H. R., KIRBY, G. W., STEGLICH, W., THOMAS, G. M., BATTERSBY, A. R., DOBSON, T. A. & RAMUZ, H. (1965). *J. Chem. Soc.*, 2423
BARTON, D. H. R., BHAKUNI, D. S., JAMES, R. & KIRBY, W. (1967). *J. Chem. Soc. C*, 128
BATTERSBY, A. R. (1967) *Pure Appl. Chem. 14*, 117
BATTERSBY, A. R. & BINKS, R. (1960) *Proc. Chem. Soc.*, 360
BATTERSBY, A. R. & HARPER, B. J. T. (1958) *Chem. Ind.*, 364
BATTERSBY, A. R. & HARPER, B. J. T. (1960) *Tetrahedron Lett.*, 21
BATTERSBY, A. R. & HERBERT, R. B. (1964) *Proc. Chem. Soc.*, 260
BATTERSBY, A. R. & MCDONALD, E. (1975) in *Falk's Porphyrins and Metalloporphyrins* (Smith, K. M., ed.), 2nd edn., Elsevier, Amsterdam
BATTERSBY, A. R., BINKS, R. & HARPER, B. J. T. (1962) *J. Chem. Soc.*, 3534
BATTERSBY, A. R., FOULKES, D. M., FRANCIS, R. J., MCCALDIN, D. J. & RAMUZ, H. (1963). *Proc. Chem. Soc.*, 203
BATTERSBY, A. R., BINKS, R., FRANCIS, R. J., MCCALDIN, D. J. & RAMUZ, H. (1964). *J. Chem. Soc.*, 3600
BATTERSBY, A. R., FOULKES, D. M. & BINKS, R. (1965*a*). *J. Chem. Soc.*, 3323
BATTERSBY, A. R., FRANCIS, R. J., RUVEDA, E. A. & STAUNTON, J. (1965*b*). *Chem. Commun.*, 89
BATTERSBY, A. R., MARTIN, J. A. & BROCHMANN-HANSSEN, E. (1967*a*) Alkaloid biosynthesis. Part X. Terminal steps in the biosynthesis of the morphine alkaloids. *J. Chem. Soc. C*, 1785–1788

BATTERSBY, A. R., FRANCIS, R. J., HIRST, M., SOUTHGATE, R. & STAUNTON, J. (1967b) Stereochemical studies concerning the biosynthesis of narcotine, protopine, and chelidonine. *Chem. Commun.*, 602–603

BATTERSBY, A. R., FOULKES, D. M., HIRST, M., PARRY, G. V. & STAUNTON, J. (1968) Alkaloid biosynthesis. Part XI. Studies related to the formation and oxidation of reticuline in morphine biosynthesis. *J. Chem. Soc. C,* 210–216

BATTERSBY, A. R., RAMAGE, R., CAMERON, A. F., HANNAWAY, C. & ŠANTAVÝ, F. (1971) 1-Phenethylisoquinoline alkaloids. Part II. The structures of alkaloids from *Colchicum cornigerum* (Sweinf.) Tackh. et Drar. *J. Chem. Soc. C*, 3514–3518

BATTERSBY, A. R., DOBSON, T. A., FOULKES, D. M. & HERBERT, R. B. (1972a) Alkaloid biosynthesis. Part XVI. Colchicine: origin of the tropolone ring and studies with the $C_6-C_3-C_6-C_1$ system. *J. Chem. Soc. Perkin Trans. I*, 1730–1735

BATTERSBY, A. R., HERBERT, R. B., PIJEWSKA, L., ŠANTAVÝ, F. & SEDMERA, P. (1972b) Alkaloid biosynthesis. Part XVII. The structure and chemistry of androcymbine. *J. Chem. Soc. Perkin Trans. I*, 1736–1740

BATTERSBY, A. R., HERBERT, R. B., MCDONALD, E., RAMAGE, R. & CLEMENTS, J. H. (1972c) Alkaloid biosynthesis. Part XVIII. Biosynthesis of colchicine from the l-phenethylisoquinoline system. *J. Chem. Soc. Perkin Trans. I*, 1741–1745

BATTERSBY, A. R., HUNT, E. & MCDONALD, E. (1973a) Biosynthesis of type-III porphyrins: nature of the rearrangement process. *J. Chem. Soc. Chem. Commun.*, 442–443

BATTERSBY, A. R., HODGSON, G. L., IHARA, M., MCDONALD, E. & SAUNDERS, J. (1973b) Biosynthesis of porphyrins and related macrocycles. Part III. Rational synthesis of [β-^{13}C]-, [γ-^{13}C]-, and [δ-^{13}C]-protoporphyrin-IX: assignment of the ^{13}C nuclear magnetic resonance signals from the *meso*-carbon atoms of protoporphyrin-IX dimethyl ester. *J. Chem. Soc. Perkin Trans. I*, 2923–2935

BATTERSBY, A. R., FRANCIS, R. J., HIRST, M., RUVEDA, E. A. & STAUNTON, J. (1975a) Biosynthesis. Part XXI. Investigations on the biosynthesis of stylopine in *Chelidonium majus*. *J. Chem. Soc. Perkin Trans. I*, 1140–1146

BATTERSBY, A. R., STAUNTON, J., WILTSHIRE, H. R., FRANCIS, R. J. & SOUTHGATE, R. (1975b) Biosynthesis. Part XXII. The origin of chelidonine and other alkaloids derived from the tetrahydroprotoberberine system. *J. Chem. Soc. Perkin Trans. I*, 1147–1155

BATTERSBY, A. R., STAUNTON, J., WILTSHIRE, H. R., BIRCHER, B. J. & FUGANTI, C. (1975c) Studies on enzyme-mediated reactions. Part V. Synthesis of (13S)- and (13R)-[13-^{3}H$_1$]scoulerine from stereospecifically labelled (R)- and (S)- [α-^{3}H$_1$]benzyl alcohol derivatives by use of liver alcohol dehydrogenase. *J. Chem. Soc. Perkin Trans. I*, 1162–1171

BATTERSBY, A. R., MCDONALD, E., REDFERN, J. R., STAUNTON, J. & WIGHTMAN, R. H. (1976a) Biosynthesis of porphyrins and related macrocycles. Part V. Structural integrity of the type-III porphyrinogen macrocycle in an active biological system; studies on the aromatisation of protoporphyrin-IX. *J. Chem. Soc. Perkin Trans. I*, 266–272

BATTERSBY, A. R., HODGSON, G. L., HUNT, E., MCDONALD, E. & SAUNDERS, J. (1976b) Biosynthesis of porphyrins and related macrocycles. Part VI. Nature of the rearrangement process leading to the natural type-III porphyrins. *J. Chem. Soc. Perkin Trans. I*, 273–282

BATTERSBY, A. R., IHARA, M., MCDONALD, E., SAUNDERS, J. & WELLS, R. J. (1976c) Biosynthesis of porphyrins and related macrocycles. Part VII. Synthesis of specifically labelled [^{14}C$_1$]uroporphyrin-III and of [10,14-^{13}C$_2$]uroporphyrin-III. Conversion of the latter into [10, 14-^{13}C$_2$]protoporphyrin-IX; biosynthetic significance of its ^{13}C nuclear magnetic resonance spectrum. *J. Chem. Soc. Perkin Trans. I*, 283–290

BATTERSBY, A. R., BUCKLEY, D. G., HODGSON, G. L., MARKWELL, R. E. & MCDONALD, E. (1976d) in *High Pressure Liquid Chromatography in Clinical Chemistry* (Nixon P. F., Gray, C. H., Lim, C. K. & Stoll, M. S., eds.), p. 63, Academic Press, London

BATTERSBY, A. R., MCDONALD, E., WILLIAMS, D. C. & WURZIGER, H. K. W. (1977a) Biosynthesis of the natural (type-III) porphyrins: proof that rearrangement occurs after head-to-tail bilane formation. *J. Chem. Soc. Chem. Commun.*, 113–114

BATTERSBY, A. R., BUCKLEY, D. G., MCDONALD, E. & WILLIAMS, D. C. (1977b) Enzymic formation of the type-III porphyrin macrocycle from unrearranged AP·AP pyrromethane. *J. Chem. Soc. Chem. Commun.*, 115–116

BATTERSBY, A. R., JOHNSON, D. W., MCDONALD, E. & WILLIAMS, D. C. (1977c) Mechanistic study of the enzymic incorporation of unrearranged AP·AP pyrromethane into uro'gen-III. *J. Chem. Soc. Chem. Commun.*, 117–118

DAUNER, H.-O., GUNZER, G., HEGER, I. & MÜLLER, G. (1976) Uroporphyrinogen formation—studies with 1-aminomethyl-3,8,13,18-tetra(2-carboxyethyl)-2,7,12,17-tetracarboxymethylbilinogen. *Hoppe-Zeyler's Z. Physiol. Chem. 357*, 147

ERDTMAN, H. & WACHTMEISTER, C. A. (1957) in *Festschrift für Arthur Stoll*, p. 144, Birkhauser, Basle

FRYDMAN, B. & FRYDMAN, R. B. (1975) Biosynthesis of uroporphyrinogens from porphobilinogen. *Acc. Chem. Res. 8*, 201–208

MARTIN, R. O., WARREN, M. E. & RAPOPORT, H. (1967) The biosynthesis of opium alkaloids. Reticuline as the benzyltetrahydroisoquinoline precursor of thebaine in biosynthesis with carbon-14 dioxide. *Biochemistry 6*, 2355-2363

ROBINSON, R. (1955) *The Structural Relations of Natural Products*, Clarendon Press, Oxford

SCOTT, A. I., HO, K. S., KAJIWARA, M. & TAKAHASHI, T. (1976) Biosynthesis of uroporphyrinogen III from porphobilinogen. Resolution of the enigmatic 'switch' mechanism. *J. Am. Chem. Soc. 98*, 1589–1591

TURNER, R. B. & WOODWARD, R. B. (1953) in *The Alkaloids*, vol. 3 (Manske, R. H. F. & Holmes, H. L., eds.), p. 54, Academic Press, New York

Discussion

McCapra: The enzyme has little impact on the rate of formation of uro'gen-III from the bilane (16). Could you chemically alter the conformation of the precursor in solution (as one can for simple peptides) so that formation of type-III would be favoured?

Battersby: That may be possible. Preliminary (unpublished) results of studies on the chemical ring closure of the bilane at different pHs show that at low pH much more type-III appears with smaller amounts of some of the other isomers; fragmentations seem to start, no doubt due to protonation within the chain. (Bilanes are notoriously difficult to handle, especially in acid conditions.)

Kirby: You could try trifluoroacetic acid. That would be a fundamental solvent change. We replaced an aqueous solvent by trifluoroacetic acid to try to mimic a biochemical process. In trifluoroacetic acid the enzyme-like reaction took place whereas in the aqueous system the reaction followed an entirely different course.

Kenner: Is this a solvent effect or is it due to a change in acidity? Trifluoroethanol, which is weakly acidic, might be a more practical solvent.

Kirby: It owes as much to the solvent change as to the acidity.

Raphael: Why should nature prefer type-III? Is there an innate advantage?

Battersby: That is the fourth question: what, when, how, *why?* In the normal course of events, uro'gen-III is formed as an intermediate on the pathway to the final macrocycles which the living system needs, especially protoporphyrin-IX (17) for haem and cytochromes. The protoporphyrin-IX produced from uro'gen-III has the unmodified propionic acid groups side by side. This ar-

rangement clearly differentiates the two halves of the molecule: the A–B half is clearly hydrophobic and the C–D half is strongly hydrophilic. In haemoglobin and myoglobin, the A–B part enters right into the hydrophobic pocket in the globin while the propionic acid groups protrude into the medium. This orientation effect may be the reason for nature's predilection for type-III.

Woodward: Type-III is statistically the favoured structure. If, many aeons ago, porphyrin biosynthesis were just a simple chemical process, the natural system might have become used to the statistically favoured type-III compounds, and later gone to pains to make them because they were familiar.

Breslow: Since enzymes are used for further conversions of the side chains they need a point of distinction in the system. The edges of the type-I uro'gen would give the enzyme no clue where to act whereas the type-III would.

Cornforth: Is the reaction of deaminase on porphobilinogen to give uro'gen-I faster than the reaction of deaminase plus cosynthetase to give uro'gen-III?

Battersby: We cannot say at present. We also want to know whether there is any difference in the rate of reaction on the bilane between the two enzyme systems.

Dewar: Have you tried solid acids adsorbed on a surface? If that produced a difference, it would fit in with the idea of pre-biological synthesis.

Battersby: We have not tried any solids yet. Bilanes have only recently been available in pure form.

Staunton: I wonder what would happen if you used a bilane in which ring D carried two acetate or two propionate groups.

Battersby: That would be interesting to know. We also want to make the

(18)
A = CH₂CO₂H ; P = CH₂CH₂CO₂H

bilane with the ring D substituents reversed, i.e. (18), to see what happens in this case.

Ramage: But you would be generating a chiral centre.

Battersby: Yes, and this could affect binding to the enzyme.

Golding: It would be interesting to compare the binding of the Fe(II) complexes of protoporphyrin-I and -IX to globin. That might contribute to the evolutionary question. You referred to the economy of nature's chemistry in the biosynthesis of morphine. However, there seems to be a certain lack of economy in that a hydroxy group is methylated to an *O*-methyl ether which is demethylated at a later stage. Surely there is no need for this methylation?

Battersby: Methylation apparently provides just the two phenolic groups needed for the oxidative step and perhaps the *O*-methyl groups may be regarded as nature's protecting groups (Barton & Cohen 1957). However, in recent work on hasubanonine and protostephanine (19) (Battersby *et al.* 1977) we found that the triphenol (20) is the precursor of both alkaloids; one of

its rings carries the *o*-diphenolic system. The diphenol (21), which *a priori* seemed to be the 'expected' precursor, was not incorporated when administered to the living plants. So here is an example where it seems that too much methylation prevents the coupling reaction.

Barton: We should qualify that statement because they could be methylated or protected by some other group which can be removed, such as glucose.

Battersby: But the precursor that is administered to the plant needs to have those three phenolic OH groups free.

Eschenmoser: Synthetic chemists would love to know if and how nature uses covalently bound protecting groups. The assignment of a protective function to the methyl group mentioned above would mean that the enzyme, acting on the unprotected substrate, would induce the same type of reaction but in a regiounselective way.

Barton: It doesn't get the chance because the substrate is methylated before it can be oxidized.

Eschenmoser: The term 'protective group' should only be used in cases where omission of the group leads to unspecific reaction.

Barton: Not necessarily; an enzyme might protect its substrate before it does anything else to it. We ought to isolate the enzymes that are responsible for the various steps.

Dewar: The fact that the *o*-diphenol is the precursor suggests that the first subsequent steps are oxidation to the *o*-quinone and nucleophilic ring closure.

Barton: This takes us back to Sir Robert's original work; when he did the oxidation to make the *o*-quinone he obtained completely different products. That is why the protection hypothesis was developed.

Woodward: One is on dangerous ground when one begins to talk about the economy of nature; we have no measure by which we can judge whether these are economical processes. Such statements are essentially without semantic content.

Battersby: My feeling is that, given the starting materials, one could not improve on the elegant routes to morphine and colchicine, for example.

Staunton: The economy of biosynthesis can be found in the fact that nature uses divergent syntheses so that many natural products can be formed from relatively few starting materials and biosynthetic intermediates.

Cornforth: If nature had any sense, she wouldn't make alkaloids!

Battersby: In one interesting case alkaloids *do* have a biological use. The male butterflies of the sub-family *Danaine* visit plants containing 1,2-dehydropyrrolizidine alkaloids which they then transform into dihydropyrrolizidines; these act as their courtship pheromones. One pheromone of this type has struc-

ture (22) (Edgar *et al.* 1971; Edgar & Culvenor 1975). So here we see a dependence of the butterfly on the alkaloids of the plant on which it feeds.

Woodward: It has been said that the many oxygenated terpenoid structures represent an unsuccessful attempt by nature to burn terpenoids to carbon dioxide. Metabolic pathways may continue to evolve in the future.

Breslow: The idea of the protecting group invokes purpose, and this is not clear. One obviously protects the OH groups if one's purpose is to make morphine, but why should the plant make morphine? Unless there is a purpose, these reactions are simply metabolic side-reactions.

Battersby: The alkaloid I just mentioned had a use but morphine (apart from providing interesting compounds for organic chemists) is at present a teleological mystery. But so was the pyrrolizidine alkaloid before the studies on the butterflies.

Barton: Why can't you accept the idea that nature was hectically making all kinds of compounds two or three thousand million years ago and in the absence of any evolutionary pressure to remove the making of compounds at random some syntheses will continue to exist? The plants are free to produce materials which they don't use because these materials are not deleterious to them. They are not under the same metabolic pressure to use every molecule as are animals, which don't make unnecessary compounds.

Breslow: That is consistent with the proposition that one should not worry about why the plant protects with the methyl groups since that is an irrelevant question in this case; the plant's purpose is not to make that compound.

Barton: Obviously, at some time it started to make that compound and it continued to do so.

Golding: I believe we can speak of the economy of nature after considering the many multistep biosyntheses that proceed *via* chiral intermediates (*cf.* Hanson & Rose 1975). In nearly all cases, the stereochemistry of each intermediate is that needed for the next step. It is unusual to find an enzymic reaction which produces an intermediate of incorrect configuration, so that an extra step is needed to epimerize to the right configuration for a subsequent step. An example of epimerization occurs in the conversion of methylmalonyl-CoA into succinyl-CoA (Sprecher *et al.* 1964). Methylmalonyl-CoA from carboxylation of propionyl-CoA has the (S)-configuration. B_{12}-dependent methylmalonyl-CoA mutase converts only the (R)-isomer into succinyl-CoA. Therefore, enzymic epimerization of (S)- to (R)-methylmalonyl-CoA is an essential step in the metabolism of propionate.

Woodward: In any biosynthetic process we are given at the start the fact that it works. How can you speak of economy when you don't know what to judge that process by?

Golding: Epimerases may have evolved to deal with the problem of mixtures of stereoisomers produced in primitive biosyntheses.

Ramage: Racemates are produced, for example, in the phenethylisoquinoline series, e.g. (±)-kreysigine (23) (Battersby *et al.* 1967), and one can get epi-

(23) (24)

merization α to the nitrogen atom through the oxidation–reduction. So nature is not so clever in that sense.

Birch: With regard to methylation: flavonoids contain the usual C_6–C_3 pattern of oxygen atoms but many flavonoids have extra introduced oxygens. These introduced oxygens are much more often methylated than those in the theoretical biogenetic precursors. It looks as if nature has introduced an OH group, then changed its mind and covered it up.

Barton: Can phenol groups not be *o*-methoxylated in one step *in vivo*?

Birch: I know of no example of that.

Ramage: The methylation of the OH in the morphine biosynthesis may prevent ring closure via *para–para* coupling which, if it happened, would preclude subsequent ring closure from the oxygen to form the five-membered ring. Moreover, demethylation might be a prelude to further metabolism *in vivo,* for instance to compounds such as acutaminine (24) in which the top ring was oxidized (Okamoto *et al.* 1969).

Battersby: We should keep in mind that the substrate for the phenol oxidation is almost certainly held on an enzyme.

Ramage: In Frank's work on phenol couplings in which the reagent was manganese dioxide on silica the phenol oxidation of reticuline went the wrong way to give isosalutaridine (Franck *et al.* 1967).

Staunton: As an enzyme is potentially a highly selective reagent I should have thought that the use of methyl protecting groups would be unnecessary *in vivo* in these phenol oxidations. Surely the enzyme could induce the required selectivity merely by holding the substrate in the appropriate conformation?

Ramage: There is a difference in the oxidation potentials of catechols and the corresponding monomethyl ethers.

Kirby: Does the phenol coupling proceed by two-site oxidation, which might go through the equivalent of a biradical, or single-site oxidation through a phenoxonium system in which two electrons would be removed from one ring and substitution occur into the other ring? The two-site process seems less 'economical' because one has to develop an enzyme with two sites which oxidize at similar rates.

Barton: I have argued against the phenoxonium ion mechanism because of its higher energy (Barton 1967).

Breslow: There is an explanation for that. If one electron is removed by oxidation of only one of the phenolic OH groups (in an enzyme system), for the coupling (with an electron acceptor still attached to oxygen) the deprotonation would have to be accompanied by concerted electron transfer (see Scheme 1).

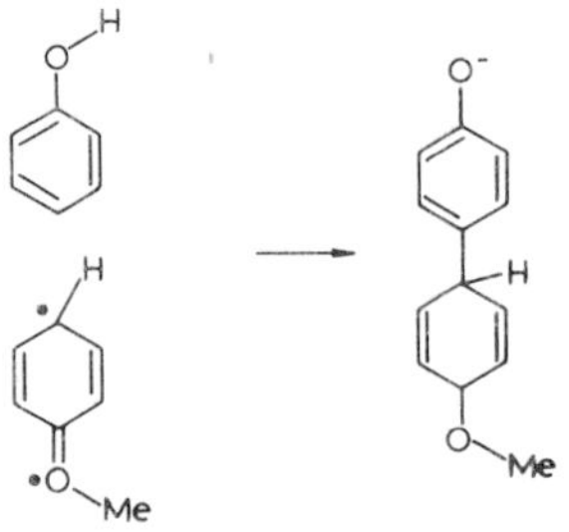

SCHEME 1 (Breslow)

That is the equivalent to attack by a phenoxonium ion but without formation of that species. Coupling of the radical with the anion would be the other one-step method of joining the rings, but that requires prior oxidation of the OH.

Dewar: But radicals are likely to be electrophilic. If a radical substitutes an anion, the electron is effectively transferred onto what was the radical. Thus initial oxidation gives the radical (25) which then attacks the anion (26). Electron transfer in the coupling reaction converts (26) into a neutral dienone and

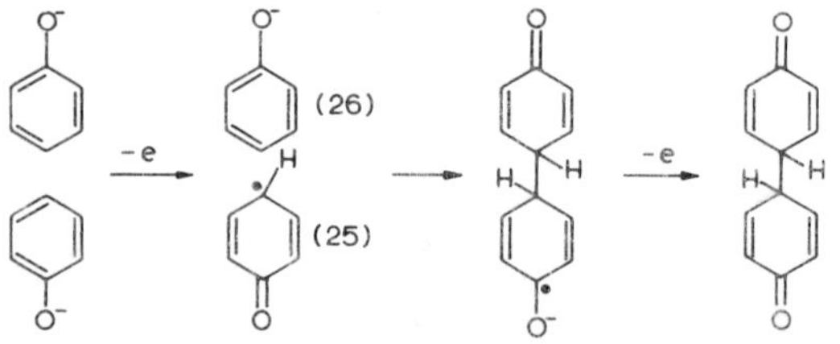

SCHEME 2 (Dewar)

(25) into a radical anion which is then oxidized (Scheme 2). Both the electrons are taken off the same oxygen atom but in two successive steps. One is removed, the anion couples, and then the second electron is removed.

Staunton: The oxidative coupling might take place by a mechanism in which oxidation gives a monooxygenated intermediate rather than by one in which electrons are abstracted to give a biradical or phenoxonium ion intermediate. For example, one ring might be converted into an arene oxide that would then be susceptible to nucleophilic attack which would give a coupling product as shown in Scheme 3. In common with the phenoxonium ion mechanism this requires oxidative attack on only one of the aryl rings.

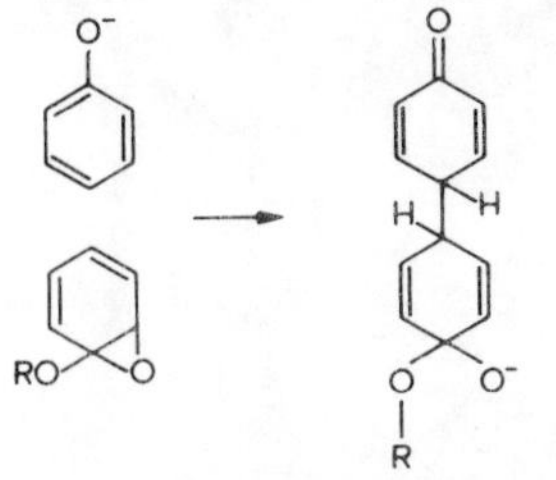

SCHEME 3 (Staunton)

Jerina *et al.* (1968) have published evidence that arene oxides are formed as intermediates in the hydroxylation of aryl rings. On this basis, therefore, the oxidative coupling of phenols is related mechanistically to that important biological process.

Ramage: But does that not contravene the 'economy of nature' in as much as these arene oxides are precursors of the phenol?

Staunton: In my view, the economy of nature lies not so much in a synthesis composed of a few steps but in the ability of a few reactions to do many different jobs.

Battersby: Why is a free hydroxy group needed on each ring?

Staunton: Perhaps the ring which undergoes oxygenation need not be phenolic. There is, however, an alternative variant of the oxygenation process in which the phenolic hydroxy group participates directly. Hamilton (1969) has suggested that hydroxylations in both aryl rings and at saturated carbon involve the generation of an activated oxygen species which can be represented formally as an oxygen atom (oxene); this intermediate is equivalent in reactivity to a carbene and so can either add an oxygen atom to a π-bond to generate an epoxide as in the generation of an arene oxide or insert an oxygen atom into a σ-bond as in the hydroxylation of a C–H bond at sp^3 carbon. If in attacking the aryl ring the oxene were to insert into one of the σ-bonds of the OH group rather than adding to a π-bond, a peroxy intermediate (27) would result which

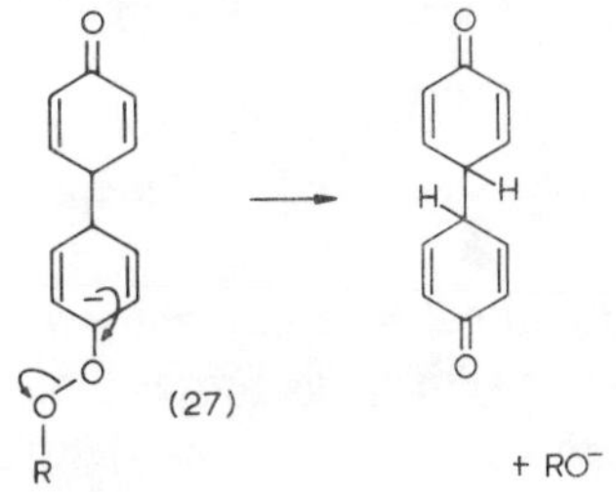

SCHEME 4 (Staunton)

could undergo a coupling reaction as shown in Scheme 4. In this case nucleophilic attack would have to take place *ortho* or *para* to the site of oxygenation.

Barton: How then do you interpret morphine biosynthesis *via* an arene oxide?

Staunton: It could be as shown in Scheme 5.

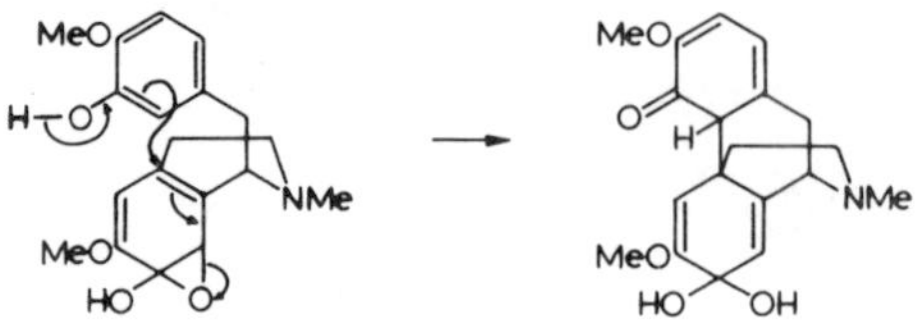

SCHEME 5 (Staunton)

Barton: Why should there be a phenolic OH in the bottom ring of the oxide? Why couldn't there be H or OMe?

Staunton: I agree that in all examples of the oxidative coupling of phenols which have been studied so far both aryl rings have had a free OH group *ortho* or *para* to the site of coupling. However this is circumstantial evidence and it does not prove that both hydroxy groups participate directly in the coupling process *in vivo;* examples may yet be discovered in which oxidative coupling takes place on a ring which lacks a free hydroxy group. As things stand the arene oxide mechanism is a reasonable alternative to the biradical mechanism for the oxidative coupling of phenols.

Barton: Many alternative mechanisms are equivalent to biradical coupling; it is just a question of *ortho–para* specificity. We know what happens *in vitro* and the results of *in vivo* experiments all fit the phenolate radical coupling hypothesis.

Staunton: The two oxygenation mechanisms differ from the various mechanisms which involve abstraction of electrons to give radical or phenoxonium ion intermediates in one important respect: it is possible that the oxygen of the phenolic hydroxy group could be replaced by one derived from molecular oxygen. It would be interesting to test this possibility by ^{18}O labelling when the enzymes become available.

Kirby: This represents another example of economy. Opening of the arene oxide gives a dienol of the type that Professor Battersby suggested was the precursor of, for example, the protostephanine system.

Woodward: It may be appropriate at this stage to quote the comment attributed to Sir Derek (Gordon Research Conference 1954): 'It pays to speculate as widely and wildly as possible; people only remember when you are right!'

References

BATTERSBY, A. R., BRADBURY, R. B., HERBERT, R. B., MUNRO, M. H. G. & RAMAGE, R. (1967) Structure and synthesis of homoaporphines: a new group of l-phenethylisoquinoline alkaloids. *Chem. Commun.*, 450–451

BATTERSBY, A. R., MINTA, A., OTTRIDGE, A. P. & STAUNTON, J. (1977) Biosynthetic derivation of hasubanonine and protostephanine from the l-benzylisoquinoline system. *Tetrahedron Lett.*, 1321–1324

BARTON, D. H. R. (1967) A region of biosynthesis. *Chem. Br.*, 330–337

BARTON, D. H. R. & COHEN, T. (1957) in *Festschrift für Arthur Stoll*, p. 117, Birkhauser, Basle

EDGAR, J. A. & CULVENOR, C. C. J. (1975) Pyrrolizidine alkaloids in *Parsonsia* species (family Apocynaceae) which attract Danaid butterflies. *Experientia 31*, 393–394

EDGAR, J. A., CULVENOR, C. C. J. & SMITH, L. W. (1971) Dihydropyrrolizine derivatives in the 'hairpencil' secretions of Danaid butterflies. *Experientia 27*, 761–762

FRANCK, B., DUNKELMANN, G. & LUBS, H. J. (1967) Synthese eines Morphinan-Derivates durch oxidativen Ringschluß. *Angew. Chem. 79*, 1066

HAMILTON, G. A. (1969) Mechanism of two- and four-electron oxidations catalysed by some metalloenzymes. *Adv. Enzymol. 32*, 55

HANSON, K. R. & ROSE, I. A. (1975) Interpretations of enzyme reaction stereospecificity. *Acc. Chem. Res. 8*, 1–10

JERINA, D. M., DAILY, J. W., WITKOP, B., NIRENBERG, P. Z. & UDENFRIEND, S. (1968) The role of arene oxide–oxepin systems in the metabolism of aromatic substrates. *J. Am. Chem. Soc. 90*, 6524

OKAMOTO, Y., YUGE, E., NAGAI, Y., KATSUTA, R., KISHIMOTO, A., KOBAYASHI, Y., KIKUCHI, T. & TOMITA, M. (1969) Acutuminine, a new alkaloid from the leaves of *Menispermin dauricum* DC. *Tetrahedron Lett.*, 1933–1936

SPRECHER, M., CLARK, M. J. & SPRINSON, D. B. (1964) The absolute configuration of methylmalonyl CoA and stereochemistry of the methylmalonyl CoA mutase reaction. *Biochem. Biophys. Res. Commun. 15*, 581–567

Design of a specific oxidant for phenols

SIR DEREK BARTON* and STEVEN V. LEY

Department of Chemistry, Imperial College, London

Abstract Selective *ortho*-substitution of phenols can be secured, in principle, by attachment of the substituting reagent to the phenolic hydroxy group and subsequent rearrangement of this derivative by a cyclic mechanism into the *ortho*-position. So far, only one example of this principle (the Claisen rearrangement) is well established. For *ortho*-hydroxylation, a phenolic ester of selenium(IV) should have the desired properties.

Diphenylseleninic anhydride, PhSe(=O)·O·Se(=O)Ph, has proved to be the new reagent for the application of this mechanism. Several phenols, including models of the phenolic ring A of tetracycline, gave *o*-hydroxydienones, with quinones and phenylselenated species as by-products, when treated with diphenylseleninic anhydride. When the phenols were converted into their corresponding anions before treatment with the anhydride, the *o*-hydroxydienones were obtained in good yield free from other products arising from reaction at the *para* position.

When phenols were added to a warm solution of diphenylseleninic anhydride, they were oxidized to the *o*-quinones even when the phenols were not substituted in the *para* position.

For several years we have been engaged in work directed to the total synthesis of the antibiotic tetracycline (1) and its congeners in a synthetically interesting way. Crucial to our approach was the introduction of a 12a-hydroxy group into

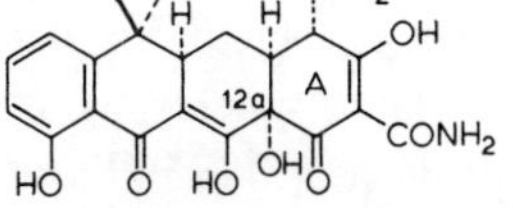

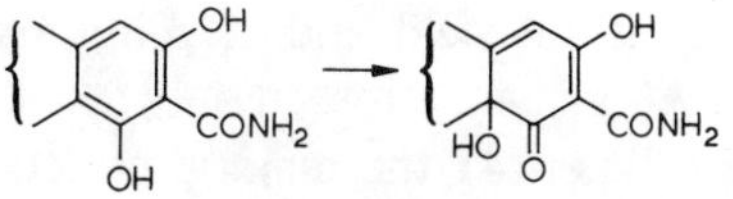

SCHEME 1

the aromatic ring A phenolic precursor (Scheme 1). Although reagents exist which will effect this transformation in simpler systems, the structural ele-

* *Present address:* Institut de Chimie de Substances Naturelles, Centre Nationale de la Recherche Scientifique, 91190 Gif-sur-Yvette, France

ments in tetracycline and ring A model phenols made their use in our synthetic strategy unsatisfactory. We needed, therefore, a new reagent which would react regiospecifically with the oxygen atom of the phenolic group (or phenolate), be able to deliver oxygen into an *ortho* position, and also be compatible with other sensitive functional groups.

Our initial attempts (Barton *et al.* 1971, 1975) with acyl peroxides, peracids or with metal oxides and hydrogen peroxide were only partially successful. However, diphenylseleninic anhydride, PhSe(=O)·O·Se(=O)Ph, which was first prepared by Doughty (1909), appeared to have many of the features that we wanted for our specific oxidizing reagent, including potential high oxidizing power, a good leaving group, PhSe(=O)·O$^-$, a reasonable solubility in organic solvents, and ease of preparation.

Surprisingly the chemistry of diphenylseleninic anhydride had not been investigated before our work. In this paper we shall discuss its reaction with simple phenols and tetracycline ring A model phenols.

Diphenylseleninic anhydride is an easily handled white solid which hydrolyses slowly in a moist atmosphere. It can be conveniently prepared by either ozonolysis of diphenyl diselenide (Ayrey *et al.* 1962) or heating *in vacuo* the nitric acid complex of phenylseleninic acid, PhSeO·OH–HNO$_3$ (Stoecker & Krafft 1906), itself formed by oxidation of diphenyl diselenide with nitric acid.

Although there is good evidence (Paetzold *et al.* 1967) that diphenylseleninic anhydride has a symmetrical structure rather than the alternative PhSeO·SeO$_2$Ph, we studied the ^{13}C n.m.r. spectrum to confirm the symmetrical nature of the molecule. The ^{13}C resonances of the phenyl ring appear at δ(TMS) 149.25 (C$_1$), 125.91 (C$_2$,C$_6$), 129.17 (C$_3$,C$_5$), and 131.95 p.p.m. (C$_4$). Had the molecule been unsymmetrical, two different sets of phenyl group carbon atoms would have been observed.

Early experiments with the anhydride and phenols showed that *ortho*-hydroxylation was achieved. However, the reaction was not regiospecific and significant amounts of the *para*-isomer were also formed together with other side products. For example, treatment of 2,4-xylenol with diphenylseleninic anhydride at room temperature gave the *o*-hydroxydienone, which was isolated as its dimer (2) in 40% yield, and the *p*-hydroxydienone (3) in 15% yield. Similar treatment of 2,6-xylenol afforded only 5% of the *o*-hydroxylated species, again isolated as its dimer (4), the major products being the quinone (5) (25%) and the biphenylquinone (6) (40%). Treatment of mesitol with the anhydride gave the *o*-hydroxydienone dimer (48%) and the *p*-hydroxydienone (30%).

Oxidation of the tetracycline ring A model phenols (7; R = OMe and R = NH$_2$) with diphenylseleninic anhydride, although giving the desired *ortho*-hydroxylated species, also gave several other products. For example, oxidation of the ester (7; R = OMe) with the anhydride at room temperature in dichloromethane afforded the o-hydroxydienone (8; R = OMe) in moderate yield and also the phenylseleno derivative (9; R = OMe). In a comparable reaction, the amide (7; R = NH$_2$) was converted into the o-hydroxydienone (8; R = NH$_2$), the phenylseleno derivative (9; R = NH$_2$) and the quinone (10; R = NH$_2$) in yields of 24, 45, and 20%, respectively.

Attempts to o-hydroxylate the ester (7; R = OMe) with Adler's sodium periodate procedure or with selenium dioxide failed.

Attempted hydroxylation with the anhydride of other ring A model phenols (11) containing a *para*-hydroxy group, however, failed. Both the models (11; R=OMe) and (11; R=NH$_2$) rapidly produced the quinone (10) on treatment with diphenylseleninic anhydride in high yield. In an effort to protect these triphenols (11; R = OMe or R = NH$_2$) from oxidation to the quinones (10) we converted them into their corresponding cyclic o-carbonates and cyclic phenyl boronates. However, treatment of these derivatives with anhydride also lead to the quinones (10).

To test whether our oxidation procedure could be used on a phenol which contained a sensitive cathylate function, a group which has potential application in tetracycline chemistry, we prepared the compound (12) by partial hydrolysis of the dicathylate (13). The structural assignment of the cathylate (12) was based mainly on ^{13}C n.m.r. measurement and comparison with a series of model compounds.

(12) (13) (14)

When the cathylate (12) was treated with the anhydride at room temperature, we were surprised to observe no reaction. A possible explanation is that there is greater hydrogen bonding in the cathylate (12) than in the ester (7; R = OMe) thus rendering attack by the phenolic OH group on the anhydride less easy.

When, however, the cathylate (12) was first converted into its phenolate anion and subsequently treated with the anhydride at 55 °C, the o-hydroxydienone (14) was obtained in 56% yield.

The results so far indicated that greater *ortho* selectivity could be obtained by prior conversion of the phenol into the corresponding anion, thereby increasing the reactivity on oxygen. After formation of the anion of 2,4-xylenol by treatment with sodium hydride and subsequent treatment with the anhydride, the o-hydroxydienone dimer (2) was produced in 45% yield. Of more importance was the fact that the p-hydroxydienone (3) was not formed in this reaction. Similarly, the anion from 2,6-xylenol gave the o-hydroxydienone dimer (4) (44%) and only trace amounts of other products. Likewise, mesitol after formation of the anion and treatment with diphenylseleninic anhydride cleanly gave the dimer in 55% yield.

Success in these reactions with the simple phenols prompted a similar study with the tetracycline ring A model phenols. Consequently, the ester (7; R = OMe) was first treated with sodium hydride to form the anion and then with the anhydride to give the o-hydroxydienone (8; R = OMe) in good yield and only a small amount of the phenylselenated species (9; R = OMe). The required o-hydroxydienone (8; R = OMe) was isolated by extraction with aqueous sodium hydrogen carbonate followed by acidification and reextraction. (This extraction process proved to be useful for the separation of these chromatographically-labile compounds.) Oxidation of the anion from the amide (7; R=NH$_2$) with the anhydride also gave a good yield of the o-hydroxydienone (8; R=NH$_2$) with only trace amounts of the phenylseleno-amide (9; R=NH$_2$) and the quinone (10; R = NH$_2$).

These results indicate that, after prior formation of the phenolate anion, oxidation with diphenylseleninic anhydride proceeds regiospecifically to give *ortho*-hydroxylation.

Although the mechanism of these reactions may be interpreted in several ways, the fact that the oxidation of the phenolate anions proceeds with a high degree of *ortho*-selectivity suggests that our original concept is justified; that is to say, that the phenolate anion reacts through its oxygen with the anhydride to produce a species which can then deliver an oxygen group specifically to the *ortho* position by a 2,3-sigmatropic rearrangement (see Scheme 2).

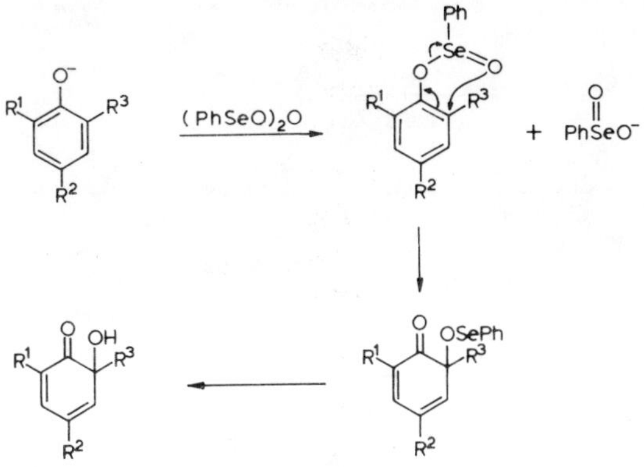

SCHEME 2

The intermediate selenic ester (15) could be responsible for phenylselenation of unreacted phenolate or might conceivably react with the phenylseleninate anion (produced in the first step) to give PhSeO·Se(=O)Ph, which itself could be either a phenylselenating species or a hydroxylating reagent.

In principle, all three oxygen atoms of the anhydride could be used in these oxidations. However, it was found that, generally, one mol equivalent of the reagent was needed to convert all the starting phenol.

Consistent with these mechanistic arguments was the fact that phenylseleninic acid was observed as a by-product on the t.l.c. baseline. Also, diphenyl diselenide was isolated as the principal by-product in all the oxidations so far studied. Other partially reduced forms of diphylseleninic anhydride were not detected. One might expect that these reduced forms would rapidly disproportionate into the anhydride and diphenyl diselenide in agreement with the observation that they also cannot be detected during the oxidation of diphenyl diselenide with ozone to the anhydride (Ayrey 1962).

The formation of *para*-hydroxylated derivatives and other products during the oxidation of phenols with diphenylseleninic anhydride undoubtedly proceeds by a different route. The reagent in these cases may act as an ordinary electrophilic oxidant towards carbon or, maybe, an electron-transfer process initiates a radical mechanism. During the course of our study on reactions of diphenylseleninic anhydride with phenols to give hydroxydienones, we were prompted to investigate the possibility of producing *ortho*-quinones by a similar reaction (Barton et al. 1976).

Few methods have been reported for the direct oxidation of phenols, particularly when the *para* position is unblocked, to *ortho*-quinones. However, phenols were smoothly converted into *o*-quinones when they were added in solution to a warm (50 °C) mixture of the anhydride in tetrahydrofuran. Only a few phenols have been investigated so far and our results are summarized in Table 1.

TABLE 1

Products and yields of direct oxidation of phenols

Starting phenol	Quinone	Yield after recrystallization (%)
α-Naphthol	*o*-Naphthoquinone	62
β-Naphthol	*o*-Naphthoquinone	63
Carvacrol	3-Methyl-6-isopropyl-*o*-benzoquinone	60
Thymol	3-Methyl-6-isopropyl-*o*-benzoquinone	59
2,4-Di-t-butylphenol	3,5-Di-t-butyl-*o*-benzoquinone	68

Mechanistically it is thought that the mechanism of *ortho* oxidation is similar to that postulated above, followed by either loss of phenylselenol (path *a*) or aromatization followed by a second oxidation (path *b*) (Scheme 3).

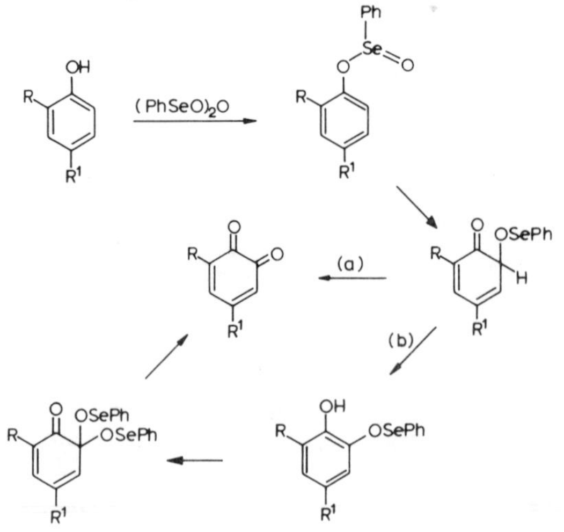

SCHEME 3

TABLE 2

Products and yields of phenols with diphenylseleninic anhydride and hexamethyldisilazane

Phenol	Product	Yield (%)
2,4-dimethylphenol	2,4-dimethyl-6-(N-phenylselenoimino)cyclohexa-2,4-dien-1-one	65
2,6-dimethylphenol	2,6-dimethyl-4-(N-phenylselenoimino)cyclohexa-2,5-dien-1-one	48
phenol	6-(N-phenylselenoimino)cyclohexa-2,4-dien-1-one	45
1-naphthol	2-(N-phenylselenoimino)naphthalen-1(2H)-one (75%) + 1-(N-phenylselenoimino)naphthalen-2(1H)-one (25%)	66
2-naphthol	2-(N-phenylselenoimino)naphthalen-1(2H)-one (40%) + 1-(N-phenylselenoimino)naphthalen-2(1H)-one (60%)	61
2-methyl-5-isopropylphenol	corresponding N-phenylselenoimino quinone	58
2-isopropyl-5-methylphenol	corresponding N-phenylselenoimino quinone	56
4-(ethoxycarbonyloxy)phenol	4-(ethoxycarbonyloxy)-6-(N-phenylselenoimino)cyclohexa-2,4-dien-1-one	45

Lastly, we have studied the oxidation of phenols by the anhydride in the presence of hexamethyldisilazane, $Me_3Si \cdot NH \cdot SiMe_3$. These reactions lead to the formation of phenylselenoimines in good yield (Table 2). Although the corresponding thioimines are known (Barton *et al.* 1973), these selenoimines represent a new class of compound. They can be readily reduced with either thiophenol, to give the aminophenol, or zinc and acetic anhydride to the corresponding amino diacetate. The method therefore constitutes a mild amination procedure for phenols.

We are currently investigating the mechanism of this interesting reaction. It seems likely that as the disilazane does not react with phenols it must react with the anhydride to give a new reactive species.

References

AYREY, G., BARNARD, D. & WOODRIDGE, D. T. (1962). *J. Chem. Soc.*, 2089

BARTON, D. H. R., MAGNUS, P. D. & PEARSON, M. J. (1971) Experiments on the synthesis of tetracycline. IX. The synthesis and rearrangement of 6-acyloxycyclohexa-2,4-dienones. *J. Chem. Soc. C,* 2231–2241

BARTON, D. H. R., BLAIR, I. A., MAGNUS, P. D. & NORRIS, R. K. (1973) The chemistry of trisulphenamides [N(SR)$_3$]. Part I. Preparation, thermal decomposition, and reactions of tribenzenesulphenamide [N(SPh)$_3$]. *J. Chem. Soc. Perkin Trans. 1,* 1031–1036

BARTON, D. H. R., MAGNUS, P. D. & QUINNEY, J. C. (1975) Experiments on the synthesis of tetracycline. XIII. Oxidation of ring A model phenols to *p*-hydroxycyclohexadienones. *J. Chem. Soc. Perkin Trans. 1,* 1610–1614

BARTON, D. H. R., BREWSTER, A. G., LEY, S. V. & ROSENFELD, M. N. (1976) Oxidation of phenols to *ortho*-quinones using diphenylseleninic anhydride. *J. Chem. Soc. Chem. Commun.*, 985

DOUGHTY, H. W. (1909). *Am. Chem. J. 41*, 326

PAETZOLD, R., BOREK, S. & WOLFRAM, E. (1967). *Z. Anorg. Chem. 353,* 53

STOECKER, M. & KRAFFT, F. (1906). *Ber. 39*, 2197

Discussion

Raphael: Do the X-ray structures show that the selenoimine is planar?

Barton: Yes; except the methyl groups, of course.

Raphael: What is the carbonyl frequency of the *o*-selenoimine (e.g. 16; X = Se) in the i.r. spectrum?

Barton: It is low: $v_{C=O}=1605$ cm^{-1}. It is more instructive to compare the carbonyl absorptions of the *o*- and *p*-compounds; $v_{C=O}$ for the *p*-compound is slightly higher, 1613 cm^{-1}.

Dewar: How does the length of the C–C bond in these selenoimines (e.g. Table 2, p. 59) and thioimines compare with that of an aromatic double bond (1.40 Å)?

Barton: For the compound (16; X = Se) and the analogous thioimine the bond lengths (in Å) are set out in Table 1.

SPECIFIC OXIDANT FOR PHENOLS

(16) X = S or Se

(17)

(18)

TABLE 1 (Barton)

Bond lengths (in Å) in the iminoquinones (16; X = Se and S)

Bond	Length	
	$X = Se$	$X = S$
C_1-C_2	1.48	1.49
C_2-C_3	1.44	1.45
C_3-C_4	1.33	1.34
C_4-C_5	1.45	1.45
C_5-C_6	1.34	1.35
C_6-C_1	1.44	1.46

Dewar: So the ring corresponds closely to the quinone. The heterocyclic form (17) is analogous to (18; X = S) which according to our calculations with MINDO/3 has strong bonding without using the d orbitals of the sulphur atoms. The short S–O bonds in the thioimine may reflect a three-centre bond formed by use of the p orbital on S, analogous to the bonding in F–Xe–F.

Barton: The S–O bond in the thioimine (17; X = S) is the same length (see Table 2) as that in the oxa-compound (18; X = O).

TABLE 2 (Barton)

Bond lengths (in Å) in the imines (17; X = S and Se)

Bond	van der Waal radius	Covalent radius	Found
S–O	3.25	1.67	2.575
Se–O	3.4	1.77	2.604

Dewar: The S–S bonds are long in the trisulphide (18; X = S) and when X = O the S–S bond shrinks and the S–O bond lengthens.

Rees: Doesn't this imply a large contribution from the ionic form (19; X = S or Se) with strong attraction between the charges? In the extreme this attraction would result in full bonding; for example, the o-diazonium phenolate anion (20) is a dipolar species but the sulphur analogue is the heterocyclic thia-

diazole (21). The bond lengths suggest that the N–Se–O five-membered heterocyclic ring has almost formed in (17). Replacement of oxygen in (17) by sulphur or selenium would probably result in formation of the heterocyclic ring with a full S–S or S–Se bond (22).

The sulphur compound (22; X = S) could be of interest since its S-oxide (23) would, I believe, provide the first example of two adjacent mono-oxidized sul-

phur atoms; normally mono-oxidation of RSO·SR gives the thiosulphone, $RSO_2 \cdot SR$, and not the disulphoxide, RSO·SOR [cf. (24)].

What might be the reactive species in the formation of selenoimines?

Barton: We favour PhSeN; some RSN species are known.

Rees: That is an attractive idea; curiously enough we tried to make the sulphur species RSN (T. A. Chaudri & C. W. Rees, unpublished work, 1971). We had generated the amino-nitrenes, $R_2N-\ddot{N}$: ↔ $R_2N^+=N^-$, which were relatively stable with some nucleophilic character, and we wanted to generalize this idea to nitrenes of the kind X–N̈: in which X has a lone pair of electrons. For X = O, oxidation of RO·NH_2 with lead tetraacetate apparently proceeds through the nitrene, reactions of which have been observed. For X = S, we hoped to generate ArS≡N ↔ $ArS^+=N^-$ ↔ ArS–N̈: by oxidation of the sulphenamide ArS·NH_2. In most cases we obtained the disulphide ArS·SAr and N_2 quantitatively and assumed that these products arose by dimerization of the transient ArSN. However, in one case we could intercept the nitrene; when the sulphenamide (25) was oxidized with lead tetraacetate in the presence of dime-

SPECIFIC OXIDANT FOR PHENOLS

O$_2$N-C$_6$H$_3$(NO$_2$)-S-NH$_2$ (25) $\xrightarrow[\text{CH}_2\text{Cl}_2; \text{Me}_2\text{SO}]{\text{Pb(OAc)}_4}$ O$_2$N-C$_6$H$_3$(NO$_2$)-S-N=S(O)(Me)Me (26)

thyl sulphoxide, the sulphoximide (26) was formed in good yield. Some inorganic compounds with the R–S≡N structure, such as FSN and ClSN, are known (Roesky 1971).

The selenium compounds RSeN may be sufficiently stable to be the reactive species that you envisage to give the selenoimines.

Prelog: What would be the best reagent to trap PhSeN?

Barton: We are using 2,6-di-t-butylphenol and trying to generate PhSeN by dehydrating the previously unknown seleninamide, PhSeO·NH$_2$. Our first efforts to make this amide gave PhSe·SePh and N$_2$.

Brown: The PhSeN could be trapped by low-valent Pt complexes, just as complexes of small-ring cyclic acetylenes have been isolated.

Rees: Me$_2$SO might also be a good trap, as I indicated.

Barton: Unfortunately one cannot dehydrate amides to generate the PhSeN when Me$_2$SO is present because it reacts with the phosgene immediately.

Rees: The selenium analogues of sulphenamides (e.g. PhSeNH$_2$) might be dehydrated with Pb(OAc)$_4$ faster than the selenium would be oxidized.

Prelog: Would PhSe·SePh attack activated nitrogen (say under electrical discharge) to produce PhSeN in a metathetic reaction?

Barton: That is a good suggestion.

Breslow: Does the disilazane not react with the seleninic anhydride?

Barton: The disilazane does not appear to react with the phenol. The phenol reacts with the anhydride but the product, in most cases, does not react with the disilazane. Most *o*-quinones do not react with the mixture of the anhydride and the silazane.

Breslow: That does not exclude the possibility that an intermediate reacts.

Barton: Spectroscopic (n.m.r., i.r.) studies of mixtures of the silazane and the anhydride do not indicate any reaction. Yet when the phenol is added at room temperature, the phenylselenoiminoquinone is rapidly formed.

Breslow: It is hard to imagine a species such as PhSeN being formed reversibly, as it is so reactive.

Barton: That is a good criticism. We do not know that PhSeN is the species. Indeed, how it would be formed is not clear.

Woodward: If the reactive species were PhSeN, it is required by your facts that it be attacked by a nucleophilic centre solely at nitrogen. Are you confident that that would be the case?

Barton: No, because we are arguing that for *o*-amination the oxygen attacks the *selenium* before the Michael rearrangement! In *p*-amination, however, it goes onto nitrogen. This is not so surprising. In experiments with $S_3N_3Cl_3$ (27)

(27)   (28)   (29)

we found that when the nucleophile was oxygen it reacted on the sulphur but when it was carbon we observed reaction on nitrogen (Barton & Bubb 1977).

Raphael: We recently came across a cognate N–O compound while looking for a good route to a starting compound (29) for the synthesis of mitomycins. We reasoned that treatment of the readily obtainable triol (28) with diazomethane would give the methoxy-diol since the two OH groups *o*- and *p*- to the NO_2 are highly acidic and the *o*-OH group will be strongly hydrogen-bonded.

(30)

We obtained the light yellow diol (29) in 50% yield but the remaining 50% was a deep red, less soluble compound (30). This could be formed by attack not at the phenol but at the nitronate ion to give a compound analogous to the selenoimines. The carbonyl in (30) absorbs at 1600 cm^{-1} in the i.r. (cf. p. 60). The *m*-OH group is not methylated. The X-ray structure is being investigated. Although (29) and (30) are isomeric their solubilities in organic solvents differ considerably. How soluble are the selenoimines?

Barton: Reasonably soluble in many solvents. Does (30) not exist as its dimer?

Raphael: It is not stable enough for us to be able to determine its molecular weight.

Barton: What is the n.m.r. shift of the proton *ortho* to the imine?

Raphael: In trifluoroacetic acid the signal for the ring proton in the expected ether is at τ 2.33; for the abnormal product it comes at τ 2.55.

Todd: Professor Barton, the yields of selenoimines (Table 2) range from 45 to 66%. What else do you obtain?

Barton: It is not one single compound but we have not identified all the products yet. The minor products must, however, be formed at the same oxidation level because we observe good titrations for oxygen consumption.

Baldwin: Since you now have a method for *o*-hydroxylating phenols why haven't you made tetracycline?

Barton: We still haven't solved the problem of putting the nitrogen function onto ring A. Also, we have partly to protect the tetracycline, leaving the right OH group free. We are working on another approach in which we add the N and the O together.

Baldwin: How can the cyclic mechanism you invoked for the *o*-hydroxylation apply to the cases of *p*-hydroxyation?

Barton: Only when we use the phenolate anion do we get *o*-hydroxylation. With the phenols, both *o*- and *p*-substitution are seen. This implies that the cyclic mechanism does not apply to attack on the phenols.

Baldwin: Even when the ethoxycarbonyloxy group was lost from the quinol which was oxidized with the nitrogen system which you say was much better than using a proton?

Barton: We like to think that the cyclic mechanism operates but the mechanism of these seleno-reactions is open to discussion. There seems to be coordination of oxygen with some species which then delivers N to the *o*-position; we observe high *o*-selectivity. When the *o*-positions are blocked, we see *p*-substitution which is particularly successful when a substituent (e.g. EtO·CO·O) that can be eliminated is present. For that reason there must be an electrophilic species that brings N into the *p*-position.

Cornforth: You propose five-membered cyclic intermediates in the *o*-hydroxylation. Have you considered six-membered rings?

Barton: Yes. We tried to make the phenyl peroxyesters (31) but failed.

(31)

Golding: Does diphenylseleninic anhydride react with enolates of ketones and esters? Introduction of an OH group adjacent to an ester carbonyl group would be useful to us for the conversion of (*S*)-glutamate into (2*S*)-4-hydroxyglutamates.

Barton: In a few preliminary experiments with cholestanone we noted elimination and oxygenation.

References

BARTON, D.H.R. & BUBB, W.B. (1977) Some reactions of unsaturated sulphur–nitrogen heterocycles with nucleophilic substrates. *J. Chem. Soc. Perkin Trans. I,* 916–923

ROESKY, H.W. (1971) in *Sulfur in Organic and Inorganic Chemistry* (Senning, A., ed.), ch. 3, Marcel Dekker, New York

Synthesis of sesquiterpenoids of biogenetic importance

ROBERT RAMAGE*

The Robert Robinson Laboratories, University of Liverpool

Abstract The contributions made by Sir Robert Robinson to sesquiterpene chemistry and to the development of the biogenetic isoprene rule are discussed. Examples of the great utility of the Robinson Ring Annelation in synthetic organic chemistry are given with reference to steroid and sesquiterpene systems. Recent modifications to the original method are also mentioned.

The close relationship of the eremophilane sesquiterpenes and the spiro sesquiterpenes, which follows from the biogenetic derivation of eremophilone by Robinson, is the basis for the synthetic strategy under discussion leading to chiral spiro sesquiterpenes. This approach makes use of chiral starting material coupled with subsequent stereospecific processes leading to sesquiterpenes of stereochemical interest and complexity.

Although sesquiterpene chemistry was not a major interest of Sir Robert Robinson, his contribution in this area of natural products was fundamental to the subsequent development of structure determination and synthesis of sesquiterpenes. The structure of eremophilone (1) (Penfold & Simonsen 1939) was due to his interpretation of the experimental results and the authors acknowledge his contribution.

To reconcile the structure (1) with the isoprene rule Robinson postulated a biogenetic route involving rearrangement of the alcohol (2). This proved to be

---

* *Present address:* Department of Chemistry, University of Manchester Institute of Science and Technology, Manchester.

a cornerstone of the biogenetic isoprene rule (Ruzicka 1959) which allowed that terpene structures could be derived *in vivo* from a regular isoprenoid system by skeletal rearrangement and 1,2-methyl migration in particular. This latter process was later recognized (Woodward & Bloch 1953) in the biosynthesis of the steroid nucleus from squalene. An earlier proposal by Robinson (1934) used squalene in such a manner that 1,2-methyl migrations were unnecessary.

A major interest of Sir Robert was the construction of the steroid ring system. During the course of developing this theme he made a major contribution to the methods of synthetic organic chemistry in the form of the Robinson Ring Annelation process which is outlined in Scheme 1 and has been recently reviewed (Gawley 1976; Jung 1976).

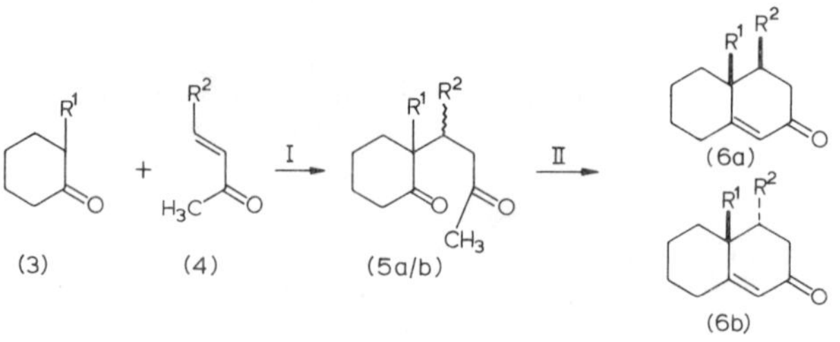

SCHEME 1

After preliminary studies with methyl vinyl ketone (Miller & Robinson 1934) and more stable enones (Rapson & Robinson 1935), it was decided to avoid methyl vinyl ketone because of its facile polymerization in the strongly basic conditions of the reaction sequence. This led to the introduction of the salt (7) (Du Feu *et al.* 1937) and the corresponding 2-chloroethyl ketone (8) (Walker 1935) as substitutes for methyl vinyl ketone. It is a tribute to the method that it was used so widely in the synthesis of the non-aromatic steroids (Woodward *et al.*

$RCH_2 \cdot CH_2 \cdot CO \cdot CH_3$ $CH_2 = CH \cdot COR$

(7) $R = \overset{+}{N}Et_2 MeI^-$ (9) $R = CH_2 \cdot CH_2 \cdot CO \cdot OCH_3$

(8) $R = Cl$ (10) $R = CH_2 \cdot CH_2 Cl$

          (14) $R = CH_2 \cdot CO \cdot OCH_3$

          (15) $R = CH_2 \cdot O \cdot CH_3$

1951, 1952; Cardwell *et al.* 1953; Sarett *et al.* 1952, 1953; Wilds *et al.* 1953; Wieland *et al.* 1953*a-c*; Johnson *et al.* 1956). In later work starting from 2-methylcyclopentane-1,3-dione the vinyl ketone (9) was developed (Velluz *et al.* 1960*a,b*) to give the potential for further cyclizations by subsequent modification of the carboxy function of (11). In a similar approach, Danishefsky & Migdalof (1969) used the enone (10) which incorporates the masked enone system of (8) and which was needed for bis-annelation of 2-methylcyclopentane-1,3-dione to give (13) *via* the intermediate (12). Bis-annelations have been useful

for the rapid construction of tricyclic and tetracyclic ring systems (Eschenmoser *et al.* 1953; Stork & McMurry 1967*b*; Stork & Ganem 1973; Ireland *et al.* 1975; Danishefsky *et al.* 1975). Although all these studies gave racemic products, recent applications of the Robinson approach to steroid synthesis using the chiral reagents have given intermediates with a high degree of optical purity (Hajos *et al.* 1968; Saucy *et al.* 1971; Saucy & Borer 1971*a, b*; Rosenberger *et al.* 1972; Eder *et al.* 1971; Hajos & Parrish 1974*a, b*).

More recently, other derivatives of methyl vinyl ketone have been introduced to extend the use of the annelation in natural product synthesis, e.g. the enone ester (14) (Pelletier *et al.* 1968; Stork & Guthikonda 1972; Trost & Kunz 1974) and the enone ether (15) (Wenkert & Berges 1967; Ireland *et al.* 1970).

The difficult step in Scheme 1 is often the initial Michael reaction, owing to slow formation of the enolate of (3) when R^1 is an alkyl substituent. One method developed for circumventing this problem was the substitution of an alkylation for the Michael reaction which led to introduction of the dichlorobutene (16) as alternative to methyl vinyl ketone (Wichterle 1947; Ireland & Kierstead 1966; Caine & Tuller 1969) and found application in steroid synthesis (Velluz *et al.* 1960*a,b*, 1961*a,b*; Burcourt *et al.* 1963). The vigorous acid hydrolysis needed to convert the intermediate vinyl chloride into an intermediate of the type (5) may be avoided by selecting the iodide (17) (Stork *et al.* 1974; Stork & Jung 1974). The reagent (18) can also be used with alkylation as the first step as it contains the methyl ketone (needed for stage II in Scheme 1) masked as a dimethylisoxazole (Stork *et al.* 1967, 1971; Stork & McMurry 1967 *a,b*). A complementary approach attempts to facilitate the Michael reaction by stabilizing the enone function of methyl vinyl ketone in the reagent (19) (Stork & Ganem 1973; Stork & Singh 1974; Boeckman 1973, 1974).

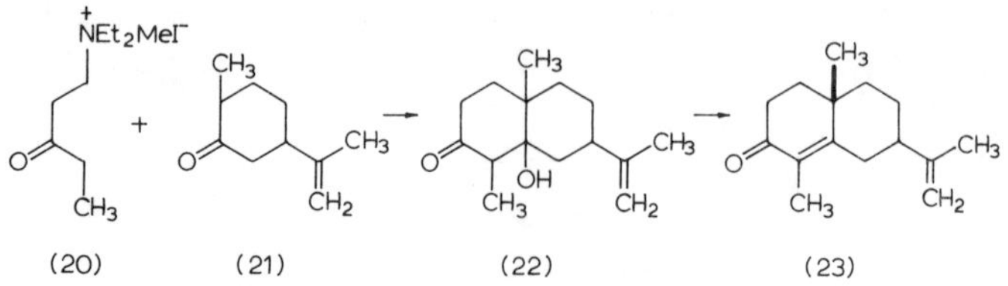

SCHEME 2

The first application of the Robinson Ring Annelation to the synthesis of a natural product was only partially successful (Adamson *et al.* 1937), being frustrated by the lack of stereochemical control in stages I and II. Sir Robert sought to synthesize α-cyperone (23) by the route shown in Scheme 2, although the nature of the ketol intermediate (22) was not elucidated until later when the stereochemical problems were solved and α-cyperone finally synthesized (McQuillin 1951, 1955; Cardwell & McQuillin 1955; Howe & McQuillin 1955). In the course of this work it was found that the use of strongly basic conditions for dehydration of the alcohol (22) also caused epimerization of the angular methyl group, possibly *via* an intermediate such as (24). Later studies of the Robinson Ring Annelation (Coates & Shaw 1968; Scanio & Starrett 1971; Marshall *et al.* 1967; Marshall & Warne 1971) have established conditions for controlling the stereochemistry of stage I to give (6a) or (6b) selectively. In the process lead-

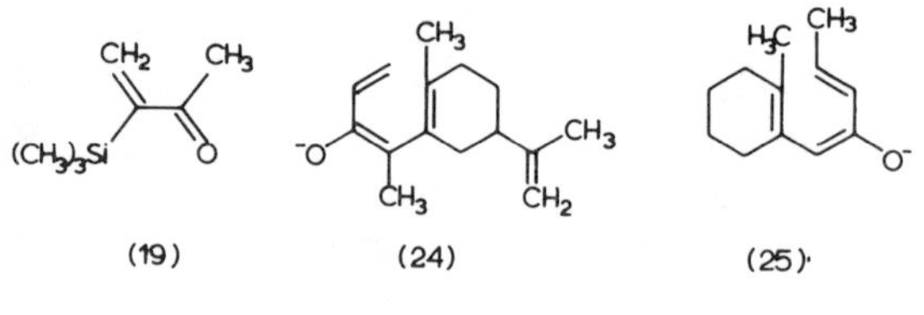

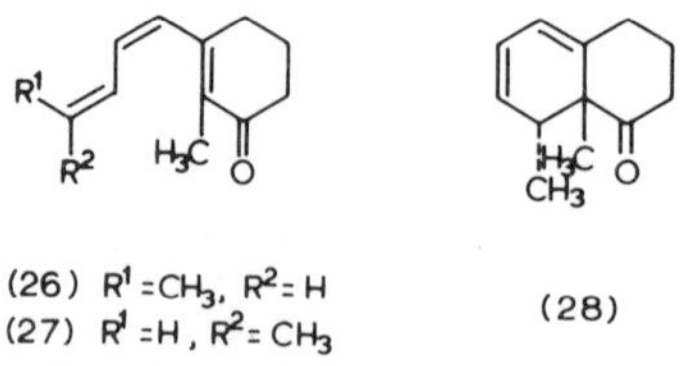

ing to (6b), where $R^1 = R^2 = CH_3$, it was postulated that the *trans*-stereochemistry results from cyclization of the intermediate triene (25) by a disrotatory process (Woodward & Hoffmann 1965, 1969). Consistent with these observations was the finding that the trienones (26) and (27) both led to the same bicyclic ketone (28) having the *trans* relationship of the two methyl groups (Ramage &

Sattar 1970). Evidently steric demands in the transition state precluded stereospecific cyclization of (26). Notwithstanding the stereochemical problems, the Robinson Ring Annelation has been the most widely used strategy for the synthesis of (±)-α-vetivone (29) and (±)-nootkatone (30) (Marshall et al. 1967; Marshall & Ruden 1970; van der Gen 1971). A recent successful approach to these important sesquiterpenes does not use this strategy (Dastur 1974).

Our interest in sesquiterpene synthesis centred on the stereochemical aspects of the biogenetic isoprene rule. In particular we were fascinated by the stereochemical relationships of the differing structural types of sesquiterpenes isolated from the commercially-important oil of vetiver. These may be exemplified by (−)-α-vetivone (29), (−)-β-vetivone (31) and (+)-zizanoic acid (32). It can readily be seen that (29) and (31) may formally represent alternative rearrangement products of a hypothetical intermediate (33), remarkably similar to (2) which was postulated by Sir Robert as a progenitor of eremophilone (1). A closely related sesquiterpene, agarospirol, had been isolated from infected agar wood and assigned the structure (34) or the enantiomer. Comparison of the structure of agarospirol with dihydroagarofuran (35) and biogenetic considerations support the absolute stereochemistry illustrated in (34). Although we had postulated (MacSweeney et al. 1970) the intermediacy of such spirosesquiterpenes in the biogenesis of zizanoic acid (32) and synthesized

(32) by a route influenced by these considerations (MacSweeney & Ramage 1971; Kido *et al.* 1969), there was an alternative hypothesis (Andersen & Syrdal 1970) which implicated the enantiomer of (+)-β-acorenol (36). Both α-acorenol (37) and β-acorenol (36) have been isolated from *Juniperus rigida*

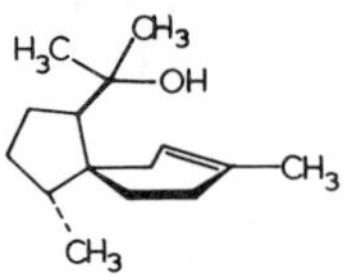

(Tomita & Hirose 1970; Tomita *et al.* 1970). The biogenetic significance of α-acorenol (37) may be illustrated by the facile *in vitro* acid-catalysed transformation of (37) into (−)-α-cedrene (38). The stereochemical complexity of spirosesquiterpenes such as (34) and (36)/(37) possessing three chiral centres, one of which is at the inaccessible spiro centre, represents a fair challenge to the synthetic organic chemist, especially if the further constraint of chirality is added to the objectives. This must necessarily be done because of the crucial stereochemical considerations involved in the biogenetic relationships of these sesquiterpenes.

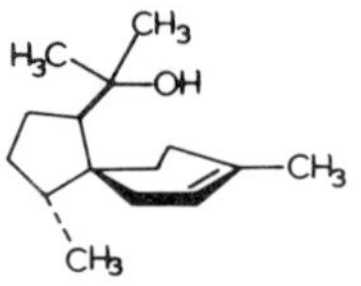

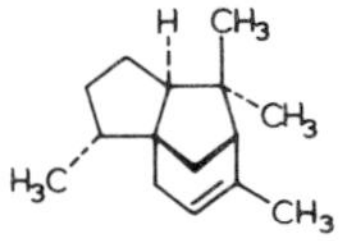

We decided to synthesize the two skeletal types—spirovetivanes and acoranes—using the strategy implicit in Scheme 3 involving the synthesis of the two chiral diastereoisomers (39) and (40). Selective ring contraction of rings *a* or *b* in the intermediates (39) and (40) should allow entry into both classes of sesquiterpenes with control over the stereochemistry at the spiro centre.

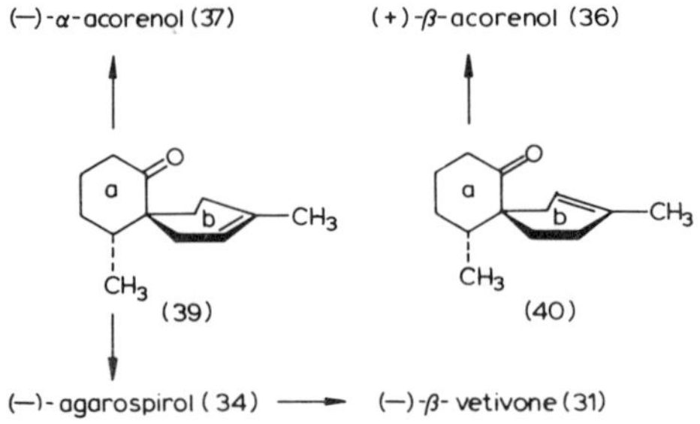

SCHEME 3

The chiral nature of the intermediates and products was dictated by the choice of starting material for the project, namely (+)-(3*R*)-methylcyclohexanone (42) which is readily available from pulegone (41). Consideration of the primary objectives (39) and (40) with this in mind meant that the spiro-annelation had to be directed to the less accessible 2-position of (42). To do this we had to protect the C-6 methylene group adjacent to the carbonyl function. During the early work on construction of the steroid nucleus, methods were devised for protection of the α-methylene before the angular methyl group is introduced (Johnson 1943; Birch & Robinson 1944; Birch et al. 1945; Johnson &

Posvic 1947). This family of protecting groups (43), (44) and (45) was later extended to include the versatile thio-analogue (46) (Ireland & Marshall 1959, 1962). Preliminary studies of the applicability of these methods led us to select the crystalline N-methylanilino derivative (47). The regiospecificity of the formylation (90%) at the 6-position was proved by oxidation to (+)-β-methyladipic acid and, furthermore, the by-product (47a) could be separated by crystallization of (47). In practice it was later found that the subsequent products derived from the small amount of (47a) could be removed at a later stage.

The approach we adopted to convert (47) into the spiro-annelated products rested on the well tried processes of cyanoethylation (Triton B/Bu^{t}OH), basic hydrolysis (Woodward et al. 1952), then esterification to give the diester (48). Dieckmann cyclization of (48) to the β-keto-ester (49) introduced, temporarily, a mixture of diastereoisomers but this complication was removed during the transformation (LiI/HCONMe$_2$) to the diketone (50). A Wittig reaction with a 10% excess of Ph$_3$P=CH$_2$ in Bu^{t}OH afforded (51) in 55% yield from (48). Reaction of (51) with ethylene glycol /p-toluenesulphonic acid gave the ketal derivatives (52) and (53) of the desired ketones (39) and (40), respectively. These diastereoisomeric ketals could be separated by chromatography on AgNO$_3$-alumina and evaluation of the c.d. and n.m.r. data allowed an assignment of stereochemistry at the spiro centre which was later to be substantiated by transformation of (52) into α-acorenol (37) and (−)-α-cedrene (38).

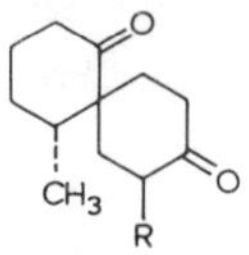

(49) R= COOCH₃
(50) R= H

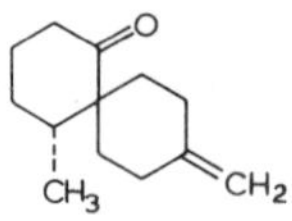

(51)

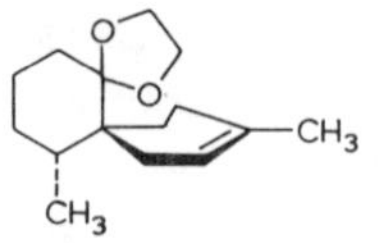

(52)

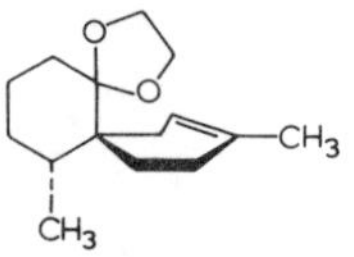

(53)

The ketal protection was cleaved in mild conditions to give (39) and (40) without interconversion through protonation of the olefinic double bond which would effectively destroy the chirality at the spiro centre. Oximation of (39) with trityllithium/pentyl nitrite gave the α-oximinoketone (54) which could be converted into the α-diazoketone (55) on treatment with chloroamine. Photolysis of (55) in alkaline solution, followed by esterification of the acidic products, afforded the ester (56) which resisted attempts to epimerize the ester

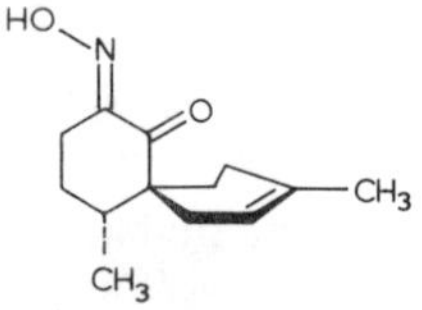

(54)

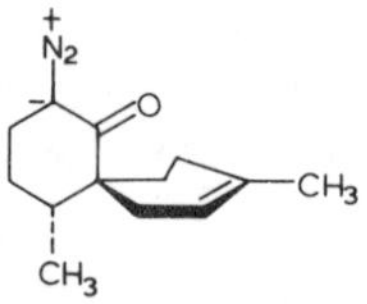

(55)

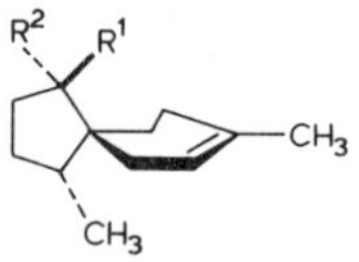

(56) R¹ = COOCH₃, R² = H
(57) R¹ = H, R² = COOCH₃

function using strong base. Consideration of conformational analysis of (56) and the epimeric ester (57) suggests the *trans* arrangement of the methyl and carbomethoxy groups to be more stable, as found in (56). This stereochemical assignment agrees with other results in this area (Corey et al. 1969; Crandall & Lawton 1969). Treatment of the ester (56) with methyllithium produced (−)-α-acorenol (37) which could be converted into (−)-α-cedrene in 96% yield with formic acid. Repetition of this synthetic route starting from (40) produced (+)-β-acorenol (36) so completing the first phase of the project (Guest et al. 1973).

Although in principle either of the ketals (52) or (53) could be used to synthesize (+)-agarospirol (34) and (−)-β-vetivone (31) we decided to proceed with the more abundant ketal (52). Ozonolysis of (52) followed by reductive cleavage of the ozonide gave the keto-aldehyde (58) which could be cyclized in basic conditions to the α,β-unsaturated ketone (59) having spectroscopic properties entirely consistent with this mode of cyclization. Hydrogenation gave the expected mixture of ketones (60) which were transformed to the corresponding acids using the bromoform reaction. Esterification followed by cleavage of the ketal protecting group gave the mixture of esters (61) and (62) which were separable by chromatography over alumina. The stereochemical assignment of the ester function in (61) and (62) followed from steric considerations and equilibration studies which showed one epimer, namely (62), to be thermodynamically more stable. The separation of (61) and (62) later proved to be unnecessary since both ketones underwent a Wittig reaction to give the same product (63) which failed to epimerize to any appreciable extent (<5%) in basic conditions. Equilibration during the Wittig reaction is to be expected in the conditions used (McMurry & von Beroldingen 1974) and assignment of the configuration of the ester function in (63) agrees with other results in this area (Yamada et al. 1973). Treatment of (63) in carefully controlled acidic conditions produced the more stable endocyclic olefin (64) without undesired rearrangement of the intermediate carbonium ion involving the adjacent spiro centre. The last two carbon atoms were added by reaction of (64) with methyllithium to give a product having analytical data in agreement with that published for (−)-agarospirol (34) (Varma et al. 1965). This synthesis proves the absolute configuration of (−)-agarospirol (34) which is therefore related to (−)-β-vetivone (31). Following an established route (Marshall & Johnson 1970) we converted the synthetic (−)-agarospirol (34) into (−)-β-vetivone (31) which was identical with the natural sesquiterpene isolated from oil of vetiver (Deighton et al. 1975). Recently there have been two new routes to spirosesquiterpenes involving elegant new spiro-annelation methods (Dauben & Hart 1975; Trost et al. 1975) but these do not have the advantages of chiral products.

The problem we are examining at present is how to transform the ketone (40) into the eremophilane sesquiterpenes which is our final objective in this area of natural product synthesis.

ACKNOWLEDGEMENTS

I am indebted to the graduate students Abdul Sattar, Clifford Hughes and Mervyn Deighton for their efforts on this research programme.

References

ADAMSON, P. S., MCQUILLIN, F. C., ROBINSON, R. & SIMONSEN, J. L. (1937) Synthetic cyperones and their comparison with α- and β-cyperones. *J. Chem. Soc.*, 1576–1581
ANDERSEN, N. H. & SYRDAL, D. D. (1970) *Tetrahedron Lett.*, 2277
BIRCH, A. J. & ROBINSON, R. (1944) The direct introduction of angular methyl groups. *J. Chem. Soc.*, 501–502
BIRCH, A. J., JAEGER, R. & ROBINSON, R. (1945) The synthesis of substances related to the sterols. XLIV. dl-*cis*-Equilenin. *J. Chem. Soc.*, 582–586
BOECKMAN, R. K. (1973) Conjugate addition-annelation. A highly regiospecific and stereospecific synthesis of polycyclic ketones. *J. Am. Chem. Soc.* 95, 6867–6869
BOECKMAN, R. K. (1974) Regiospecificity in enolate reactions with α-silyl vinyl ketones. An application to steroid total synthesis. *J. Am. Chem. Soc.* 96, 6179–6181
BURCOURT, R., TESSIER, J. & NOMINÉ, G. (1963). *Bull. Soc. Chim. Fr.*, 1923
CAINE, D. & TULLER, F. N. (1969) An alternative synthesis of *trans*-8,10-dimethyl-1(9)-octal-2-one. *J. Org. Chem.* 34, 222–223
CARDWELL, H. M. E. & MCQUILLIN, F. J. (1955). *J. Chem. Soc.*, 525
CARDWELL, H. M. E., CORNFORTH, J. W., DUFF, S. R., HOLTERMANN, H. & ROBINSON, R. (1953) Experiments on the synthesis of substances related to the sterols. LI. Completion of the synthesis of androgenic hormones and of cholesterol group of sterols. *J. Chem. Soc.*, 361–384
COATES, R. M. & SHAW, J. E. (1968) Stereoselectivity in the synthesis of *cis*- and *trans*-4,4α-dimethyl-2-octalone derivatives. *Chem. Commun.*, 47
COREY, E., J., GIROTRA, N. N. & MATHEW, C. T. (1969) Total synthesis of *dl*-cedrene and *dl*-cedrol. *J. Am. Chem. Soc.* 91, 1557–1559
CRANDALL, T. G. & LAWTON, R. G. (1969) A biogenetic-type synthesis of cedrene. *J. Am. Chem. Soc. 91*, 2127–2129
DANISHEFSKY, S. & MIGDALOF, B. H. (1969) β-Chloroethyl vinyl ketone, a useful reagent for the facile construction of fused ring systems. *J. Am. Chem. Soc.* 91, 2806–2807
DANISHEFSKY, S., CAIN, P. & NAGEL, A. (1975) Bis annelations via 6-methyl-2-vinylpyridine. An efficient synthesis of *dl*-D-homoestrone. *J. Am. Chem. Soc.* 97, 380–387
DASTUR, K. P. (1974) A stereoselective approach to eremophilane sesquiterpenes. A synthesis of (±)-nootkatone and (±)-α-vetivone. *J. Am. Chem. Soc.* 96, 2605–2608
DAUBEN, W. G. & HART, D. J. (1975) The total synthesis of spirovetivanes. *J. Am. Chem. Soc. 97*, 1622–1623
DEIGHTON, M., HUGHES, C. R. & RAMAGE, R. (1975) Stereospecific synthesis of (−)-agarospirol and (−)-β-vetivone. *J. Chem. Soc. Chem. Commun.*, 662
DU FEU, E. C., MCQUILLIN, F. J. & ROBINSON, R. (1937) Experiments on the synthesis of substances related to sterols. XIV. A simple synthesis of certain octalones and ketotetrahydrohydrindenes which may be of angle-methyl-substituted type. A theory of the biogenesis of sterols. *J. Chem. Soc.*, 53–60

EDER, U., SAUER, G. & WIECHERT, R. (1971) Neuartige asymmetrische Cyclisierung zu optisch aktiven Steroid-CD-Teilstücken. *Angew. Chem. 83*, 492

ESCHENMOSER, A., SCHREIBER, J. & JULIA, S. A. (1953) Steroids and sex hormones. Synthesis of 8,10a-dimethyl-1,7-dioxo-$D^{4a,8}$-decahydrophenanthrene. *Helv. Chim. Acta 36,* 482

GAWLEY, R. E. (1976) Robinson annelation and related reactions. *Synthesis,* 777–794

GUEST, I. G., HUGHES, C. R., RAMAGE, R. & SATTAR, A. (1973) Stereospecific synthesis of (−)-α-acorenol and (+)-β-acorenol. *J. Chem. Soc. Chem. Commun.,* 526

HAJOS, Z. G. & PARRISH, D. R. (1974a) Synthesis and conversion of 2-methyl-2-(3-oxobutyl)-1,3-cyclopentanedione to the isomeric racemic ketols of the [3.2.1]bicyclooctane and of the perhydroindan series. *J. Org. Chem. 39*, 1612–1614

HAJOS, Z. G. & PARRISH, D. R. (1974b) Asymmetric synthesis of bicyclic intermediates of natural product chemistry. *J. Org. Chem. 39*, 1615–1621

HAJOS, Z. G., PARRISH, D. R. & OLIVETO, E. P.(1968) Total synthesis of optically active (−)-17β-hydroxy-$\Delta^{9,10}$-desA-androsten-5-one. *Tetrahedron 24*, 2039–2048

HOWE, R. & MCQUILLIN, F. J. (1955). *J. Chem. Soc.*, 2425

IRELAND, R. E. & MARSHALL, J. A. (1959). *J. Am. Chem. Soc. 81*, 6336

IRELAND, R. E. & MARSHALL, J. A. (1962). *J. Org. Chem. 27*, 1615, 1620

IRELAND, R. E. & KIERSTEAD, R. C. (1966). *J. Org. Chem. 31*, 2543

IRELAND, R. E., MARSHALL, D. R. & TILLEY, J. W. (1970) A convenient stereoselective synthesis of 9,10-dimethyl-*trans*-1-decalones through the photolysis of fused methoxycyclopropanes. *J. Am. Chem. Soc. 92*, 4754–4756

IRELAND, R. E., DAWSON, M. I.,, KOWALSKI, C. J., LIPINSKI, C. A., MARSHALL, D. R., TILLEY, J. W., BORDNER, J. & TRUS, B. L. (1975) Experiments directed towards the total synthesis of terpenoids. Synthesis of 8-methoxy-4aβ,10aβ,12aα-trimethyl-3,4,4a,4bβ,5,6,10b,11,12,12a-decahydrochrysen-1(2*H*)-one, a key intermediate in the total synthesis of (+)-shionone. *J. Org. Chem. 40*, 973–989

JOHNSON, W. S. (1943) Introduction of the angular methyl group. Preparation of *cis*- and *trans*-1-keto-9-methyldecahydronaphthalene. *J. Am. Chem. Soc. 65*, 1317–1324

JOHNSON, W. S. & POSVIC, H. (1947) Introduction of angular methyl group. Alkoxymethylene blocking. *J. Am. Chem. Soc. 69*, 1361

JOHNSON, W. S., BANNISTER, B., BLOOM, B. M., KEMP, A. D., PAPPO, R., ROGIER, E. R. & SZMUSZKOVICZ, J. (1956). *J. Am. Chem. Soc.78*, 6331

JUNG, M. E.(1976) A review of annulation. *Tetrahedron 32*, 3–32

KIDO, F., UDA, H. & YOSHIKOSHI, A. (1969) Total synthesis of zizaane-type sesquiterpenoids. *J. Chem. Soc. Chem. Commun.*, 1335

MCQUILLIN, F. J. (1951) *J. Chem. Soc.*, 716

MCQUILLIN, F. J. (1955) *J. Chem. Soc.*, 528

MCMURRY, J. E. & VON BEROLDINGEN, L. A. (1974) Ketone methylation without epimerization; total synthesis of (±)-laurene. *Tetrahedron 30*, 2027–2032

MACSWEENEY, D. F. & RAMAGE, R. (1971) A stereospecific total synthesis of zizanoic and isozizanoic acids. *Tetrahedron 27*, 1481–1490

MACSWEENEY, D. F., RAMAGE, R. & SATTAR, A. (1970) Biogenetic relationship of the vetiver sesquiterpenes. *Tetrahedron Lett.,* 557–560

MARSHALL, J. A. & JOHNSON, P. C. (1970) The structure and synthesis of β-vetivone. *J. Org. Chem. 35*, 192–195

MARSHALL, J. A. & RUDEN, R. A. (1970) The stereoselective total synthesis of racemic nootkatone. *Tetrahedron Lett.*, 1239–1242

MARSHALL, J. A. & WARNE, T. M. (1971) The total synthesis of (±)-isonootkatone. Stereochemical studies of the Robinson annelation reaction with 3-penten-2-one. *J. Org. Chem. 36*, 178–183

MARSHALL, J. A., FAUBLE, H. & WARNE, T. M. (1967) The total synthesis of racemic isonootkatone (α-vetivone). *Chem. Commun.*, 753

MILLER, S. A. & ROBINSON, R. (1934) Condensation of phenols with unsaturated ketones or alde-

hydes. I. β-Naphthol and vinyl methyl ketone. *J. Chem. Soc.*, 1535–1536
PELLETIER, S. W., CHAPPELL, R. L. & PRABHAKAR, S. (1968) A stereoselective synthesis of racemic andrographolide lactone. *J. Am. Chem. Soc. 90*, 2889–2894
PENFOLD, A. R. & SIMONSEN, J. L. (1939) Constitution of eremophilane, hydroxy- and hydroxy-dihydro-eremophilane. *J. Chem. Soc.*, 87–89
RAMAGE, R. & SATTAR, A. (1970) Thermal isomerization of a hexatriene system: synthesis and rearrangement of 2-methyl-3-(cis, cis-penta-1,3-dienyl)cyclohex-2-en-1-one. *J. Chem. Soc. Chem. Commun.*, 173
RAPSON, W. S. & ROBINSON, R. (1935) Experiments on the synthesis of substances related to sterols. II. A new general method for the synthesis of substituted *cyclo*hexenones. *J. Chem. Soc.*, 1285–1288
ROBINSON, R. (1934) Structure of cholesterol. *Chem. Ind. 53*, 1062–1063
ROSENBERGER, M., DUGGAN, A. J., BORER, R., MULLER, R. & SAUCY, G. (1972) Steroid total synthesis. (+)-Estr-4-ene-3,17-dione. *Helv. Chim. Acta 55*, 2663–2673
RUZICKA, L. (1959). *Proc. Chem. Soc.*, 341
SARETT, L. H., LUKES, R. M., BEYLER, R. E., POOS, G. L., JOHNS, W. F. & CONSTANTIN, J. M. (1953). *J. Am. Chem. Soc. 74*, 1393, 1405, 4974
SARETT, L. H., LUKES, R. M., BEYLER, R. E., POOS, G. I., JOHNS, W. F. & CONSTANTIN, J. M. (1953). *J. Am. Chem. Soc. 75*, 422, 1707, 2112
SAUCY, G. & BORER, R. (1971a) Steroid total synthesis. (–)-17-hydroxy-des-A-androst-9-en-5-one. *Helv. Chim. Acta 54*, 2121–2132
SAUCY, G. & BORER, R. (1971b) Steroid total synthesis. 3. 9,10-Testosterone. *Helv. Chim. Acta 54*, 2517–2518
SAUCY, G., BORER, R. & FÜRST, A. (1971) Total Synthese von Steroiden. *Rac*-17-Hydroxy-des-A-androst-9-en-5-on. *Helv. Chim. Acta 54*, 2034–2042
SCANIO, C. J. V. & STARRETT, R. M. (1971) A remarkably stereoselective Robinson annelation reaction. *J. Am. Chem. Soc. 93*, 1539–1540
STORK, G. & GANEM, B. (1973) α-Silylated vinyl ketones. A new class of reagents for the annelation of ketones. *J. Am. Chem. Soc. 95*, 6152–6153
STORK, G. & GUTHIKONDA, R. N. (1972) Sterioselective total synthesis of (±)-yohimbine, ψ-yohimbine, and (±)-β-yohimbine. *J. Am. Chem. Soc. 94*, 5109–5110
STORK, G. & JUNG, M. E. (1974) Vinylsilanes as carbonyl precursors. Use in annelation reactions. *J. Am. Chem. Soc. 96*, 3682–3684
STORK, G. & MCMURRY, J. E. (1967a) The mechanism of isoxazole annelation. *J. Am. Chem. Soc. 89*, 5463
STORK, G. & MCMURRY, J. E. (1967b) Stereospecific synthesis of steroids *via* isoxazole annelation. *dl*-D-Homotestosterone and *dl*-progesterone. *J. Am. Chem. Soc. 89*, 5464
STORK, G. & SINGH, J. (1974) Regiospecific Michael reactions in aprotic solvents with α-silylated electrophilic olefins. Application to annelation reactions. *J. Am. Chem. Soc. 96*, 6181–6182
STORK, G., DANISHEFSKY, S. & OHASHI, M. (1967). *J. Am. Chem. Soc. 89*, 5459
STORK, G., OHASHI, M., KAMACHI, H. & KAKISAWA, H. (1971) A new pyridine synthesis *via* 4-(3-oxoalkyl)isoxazoles. *J. Org. Chem. 36*, 2784–2786
STORK, G., JUNG, M. E., COLVIN, E. & NOEL, Y. (1974) Synthetic route to halomethyl vinylsilanes. *J. Am. Chem. Soc. 96*, 3684–3686
TOMITA, B. & HIROSE, Y. (1970) Terpenoids. XXVI. Acoradiene and acorenol, key intermediates of cedrane group sesquiterpenoids, and their transformation into (–)-α-cedrene. *Tetrahedron Lett.*, 143–144
TOMITA, B., ISONO, T. & HIROSE, Y. (1970). Terpenoids. XXVIII. Acorane type sesquiterpenoids from *Juniperus rigida* and hypothesis for the formation of new tricarboxylic sesquiterpenoids. *Tetrahedron Lett.*, 1371–1372
TROST, B. M. & KUNZ, R. A. (1974) New synthetic reactions. A convenient approach to methyl 3-oxo-4-pentenoate. *J. Org. Chem. 39*, 2648–2649

TROST, B. M., HUROI, K. & HOLY, N. (1975) A new stereocontrolled approach to spirosesquiterpenes. Synthesis of acorenone B. *J. Am. Chem. Soc. 97*, 5873–5877
VAN DER GEN, A., VAN DER LINE, L. M., WITTEVEEN, T. G. & BOELENS, H. (1971) Stereoselective synthesis of eremophilane sesquiterpenoids from β-pinene. *Rec. Trav. Chim. 90*, 1034
VARMA, K. R., MAKESHWARI, M. L. & BHATTACHARYYA, S. C. (1965). *Tetrahedron 21*, 115
VELLUZ, L., NOMINÉ, G. & MATHIEU, J. (1960a). *Angew. Chem. 72*, 725
VELLUZ, L., NOMINÉ, G., MATHIEU, J., TOROMANOFF, E., BERTIN, D., TISSIER, J. & PIERDET, A. (1960b). *Compt. Rend. Acad. Sci. Paris 250*, 1084
VELLUZ, L., NOMINÉ, G., BURCOURT, R., PIERDET, A. & DUFAY, PH. (1961a). *Tetrahedron Lett.*, 127
VELLUZ, L., NOMINÉ, G., BURCOURT, R., PIERDET, A. & TISSIER, J. (1961b). *Compt. Rend. Acad. Sci. Paris 252*, 3903
WALKER, J. (1935). *J. Chem. Soc.*, 1585
WENKERT, E. & BERGES, D. A. (1967). *J. Am. Chem. Soc. 89*, 2507
WICHTERLE, O. (1947) Transformation of vinyl chlorides to ketones. *Coll. Czech. Chem. Commun. 12*, 93–100
WIELAND, P., UEBERWASSER, H., ANNER, G. & MIESCHER, K. (1953a) Steroids. The preparation of 8,10a-dimethyl-1,7-dioxo-Δ^8-dodecahydrophenanthrene. *Helv. Chim. Acta 36*, 376–386
WIELAND, P., ANNER, G. & MIESCHER, K. (1953b) [Steroids. The steric relationship of $\Delta^{8(8a)}$-1,7-dioxo-8,10a-dimethyldodecahydrophenanthrene with the sterols. Total syntheses in the sterol series II.] *Helv. Chim. Acta 36*, 646–652 (Ger)
WIELAND, P., UEBERWASSER, H., ANNER G., & MIESCHER, K. (1953c) [Steroids. Total synthesis of D-homosteroids]. *Helv. Chim. Acta 36*, 1231–1241 (Ger)
WILDS, A. L., RALLS, J. W., TYNER, D. A., DANIELS, R., KRAYCHY, S. & HARNICK, M. (1953) Total synthesis of racemic methyl 3-oxoetiocholanate. *J. Am. Chem. Soc. 75*, 4878–4879
WOODWARD, R. B. & BLOCH, K. (1953) Cyclization of squalene in cholesterol synthesis. *J. Am. Chem. Soc. 75*, 2023
WOODWARD, R. B. & HOFFMANN, R. (1965). *J. Am. Chem. Soc. 87*, 395
WOODWARD, R. B. & HOFFMANN, R. (1969). *Angew. Chem. 81*, 797
WOODWARD, R. B., SONDHEIMER, F., TAUB, D., HEUSLER, K. & MCLAMORE, W. M. (1951) Total synthesis of a steroid. *J. Am. Chem. Soc. 73*, 2403–2404
WOODWARD, R. B., SONDHEIMER, F., TAUB, D., HEUSLER, K. & MCLAMORE, W. (1952) The total synthesis of steroids. *J. Am. Chem. Soc. 74*, 4243–4251
YAMADA, K., NAGASE, H., HAYAKAWA, Y., AOKI, K. & HIRATA, Y. (1973) Synthetic studies on spirovetivanes. 1. Spirocondensation of a 4-(3-formylpropyl)-3-cyclohexenone and stereospecific total synthesis of *dl*-β-vetivone. *Tetrahedron Lett.*, 4963–4966

Discussion

Baker: We have isolated two new sesquiterpenes (65) and (66) from the first defensive secretions of termites. The first, (65) (from *Amitermes evuncifer*) (Wadhams *et al.* 1974), is an analogue of dihydroagarofuran (cf. [35]) and has been synthesized by an unambiguous route from (–)-carvone (Scheme 1) (Baker *et al.* 1977). The annelation and the ring closure by oxymercuration are the important steps and yield a product ($[\alpha]_D^{28}$ –22°), identical with the naturally occurring material.

We were interested in this compound because of its insecticidal properties; the termites protect themselves by squirting this material at their predators,

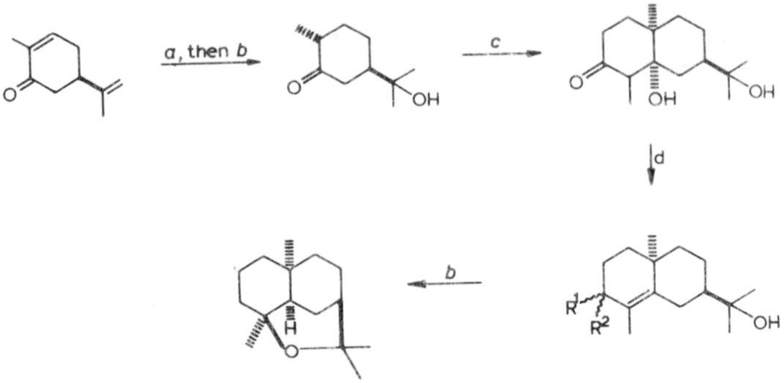

SCHEME 1 (Baker): *a*, Zn–NaOH–EtOH–H$_2$O; *b*, (i) Hg(OAc)$_2$–THF–H$_2$O, (ii) alkaline NaBH$_4$; *c*, (i) NaH–THF, (ii) ClCH$_2$·CH$_2$·CO·CH$_2$·CH$_3$; *d*, (i) conc. HCl–EtOH, (ii) LiAlH$_4$–THF, (iii) Ac$_2$O–pyridine, (iv) Li–liq. NH$_3$: (i) R^1, R^2 = O; (ii) R^1 = H, R^2 = OH; (iii) R^1 = H, R^2 = OAc; (iv) R^1 = R^2 = H.

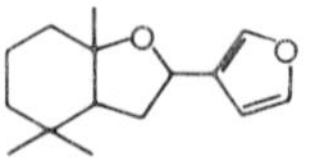

ants. We do not yet know how insecticidal it is or the origin of its toxicity.

The other sesquiterpene (66) is also a termite defensive secretion (from *Ancistrotermes cavithorax*) and is possibly used for the protection against predators. Again we do not know how insecticidal it is and have not yet completed our synthesis of this material. We wonder whether insects accumulate within themselves selective insecticides.

Barton: Dr Ramage, cyclization of (67) (cf. [58], p. 74) forms a stable ketol (68). Can you explain that?

Ramage: Scheme 2 depicts what we think happens.

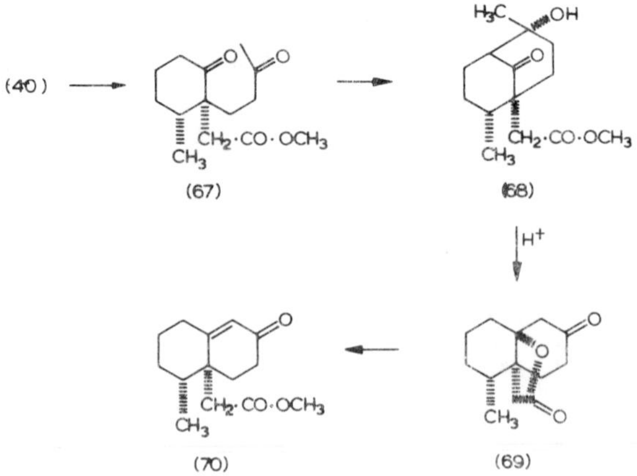

SCHEME 2 (Ramage)

Barton: Similar reactions often occur in the synthesis of the AB rings of steroids. Usually treatment of the ketol with NaOMe or NaOH in EtOH/H$_2$O gives the αβ-unsaturated ketone.

Ramage: Acid treatment of (68) gives what we believe is the lactone (69).

Barton: Treatment of that with base and methyl iodide should give the methyl ester.

Ramage: My target is the enone (70), in which the carbonyl carbon atom of the ester will form the basis of the isopropyl group in the 3-position (α- to the ketone). We intend to generate a tricyclic system and then obtain the methyl group by ring fission. In this way we can usefully introduce another chiral centre.

Birch: Dastur (1974) has implemented my suggestion for producing the *cis*-methyl groups. In Diels–Alder reactions on methylcyclohexadienes (see Scheme 3) the acrylic ester obviously attacks from the opposite side to the

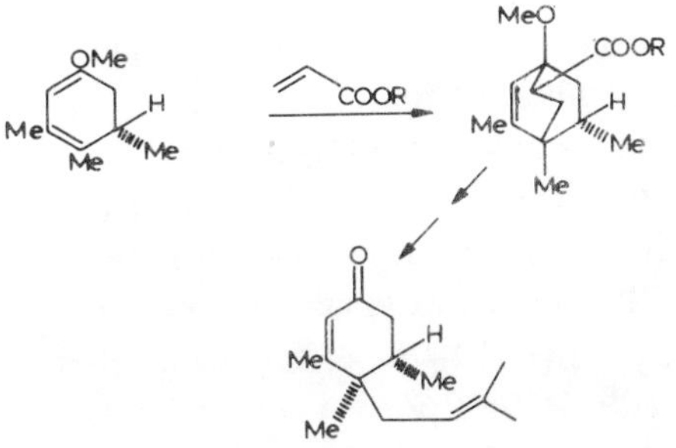

SCHEME 3 (Birch)

asymmetric methyl group. The adduct can be converted into a tertiary alcohol by a Grignard reaction. On treatment of the alcohol with acid, the ring opens to give the αβ-unsaturated ketone with two *cis*-methyl groups. To synthesize α-vetivone and nootkatone Dastur finally closed the required ring. This method is generally applicable.

Ramage: That is probably the best synthesis of the racemic eremophilane system to date, in terms of yield.

Jones: In 1935 we succeeded by classical methods in making a sesquiterpene, dihydrocyperone, the first with an angular methyl group. Sir Robert, however, stole our thunder by telling us that he had just the method for making cyperone itself by his ring annelation, and his group later synthesized it (Adamson *et al.* 1937).

A few years later, Sir Robert told Heilbron about δ-hexenolactone which had been isolated from mountain ash and which showed differential inhibitory properties on animal tissues (Medawar *et al.* 1943). Sir Robert had been trying to synthethize this for some time without any success. I suggested that we could make it by taking advantage of our knowledge of the use of acetylenic compounds in synthesis and, on Heilbron's advice, I wrote to Sir Robert. He replied expressing his interest and ended the letter by saying that he would 'be happy to test the product you obtain, if any'. Two weeks later he had 10 g (Haynes & Jones 1946).

Ramage: Sir Robert must have been inspired by the ring annelation because he subsequently produced a paper on steroid biosynthesis based on the annelation. He must, therefore, have disregarded his previous ideas about squalene.

Birch: I once congratulated Sir Robert on deriving steroids from squalene whereupon he laughed and replied that since then he had six other ideas and all of them could not be right. In fact, his route from squalene was incorrect in detail.

Todd: Sir Robert once confided to me that 'we all make mistakes; but I don't make stupid mistakes'.

Prelog: During World War II, before I became a stereochemist, Ruzicka suggested that I continue to work on alkaloids, but the only alkaloid available to me at that time was strychnine. Remembering my teacher in Prague, Votoček, who used to say that a newborn calf is not afraid of tigers, I started work on it. By that time Hermann Leuchs had written 122 papers, Sir Robert 41, and Heinrich Wieland 30 about the structure of *Strychnos* alkaloids. Sir Robert had deduced for strychnine the formula (71) which was generally accepted. We were lucky to discover that this formula cannot be correct and we rashly proposed formula (72) (Prelog & Szpilfogel 1945). Sir Robert was irritated that we changed his formula and wrote (1946): 'If Prelog and Szpilfogel are able to provide new evidence that ring E is six-membered the formula that must be considered is III [= 73].'

(71) (72) (73)

After that it occurred to him that *Strychnos* and *Cinchona* alkaloids are biogenetically related, and he published (1947) the unlikely formula (74) for strychnine. This may be a salutory example of how too much reliance on biogenetic principles can lead to error.

The correctness of 'formula III' (73) was soon proven by Professor Woodward by a brilliant analysis of all known facts (Woodward & Brehm 1948).

When we found (Goutarel *et al.* 1950) that cinchonamine (75) is the missing

(74)   (75)

biogenetic link between *Strychnos* and *Cinchona* alkaloids, Sir Robert was pleased because this showed that although he was wrong he had *not* made a 'stupid' mistake!

References

ADAMSON, P. S., MCQUILLIN, F. C., ROBINSON, R. R. & SIMONSEN, J. L. (1937) Synthetic cyperones and their comparison with α- and β-cyperones. *J. Chem. Soc.*, 1576–1581
BAKER, R., EVANS, D. A. & MCDOWELL, P. G. (1977) Stereospecific synthesis of 4,11-epoxy-*cis*-eudesmane, a tricyclic sesquiterpene defence secretion from the termite *Amitermes evuncifer*. *J. Chem. Soc. Chem. Commun.*, 111
DASTUR, K. P. (1974) A stereoselective approach to eremophilane sesquiterpenes. A synthesis of (±)-nootkatone and (±)-α-vetivone. *J. Am. Chem. Soc. 96*, 2605–2608
GOUTAREL, R., JANOT, M.-M., PRELOG, V. & TAYLOR, W. I. (1950) Cinchona alkaloids. VII. The construction of cinchonine and quinamine. *Helv. Chim. Acta 33*, 150–164
HAYNES, L. J. & JONES, E. R. H. (1946) Unsaturated lactones. I. (Acetylenic compounds. X). New route to growth-inhibitory αβ-ethylenic γ- and δ-lactones. *J. Chem. Soc.*, 954–957
MEDAWAR, P. B., ROBINSON, G. M. & ROBINSON, R. (1943) A synthetic differential growth inhibitor. *Nature (Lond.) 151*, 195
PRELOG, V. & SZPILFOGEL, S. (1945) Constitution of strychnine. *Experientia 1*, 197–198
ROBINSON, R. (1946) Constitution of strychnine. *Experientia 2*, 28–29
ROBINSON, R. (1947) Strychnine and its relation to cinchonine. *Nature (Lond.) 159*, 263
WADHAMS, L. J., BAKER, R. & HOWSE, P. E. (1974) 4,11-Epoxy-*cis*-eudesmane, a novel oxygenated sesquiterpene in frontal gland secretion of termite, *Amitermes evuncifer* Silvestri. *Tetrahedron Lett.*, 1697
WOODWARD, R. B. & BREHM, W. J. (1948) Structure of strychnine. Formulation of the *neo*-bases. *J. Am. Chem. Soc. 70*, 2107–2115

Rules for ring closure

J. BALDWIN

Chemistry Department, Massachusetts Institute of Technology, Cambridge, Massachusetts

Abstract The ease of formation of cyclic structures shows some regularity which can be described by a set of 'rules' for ring closure. The restrictions on the endocyclic transfer of groups due to S_N2 displacements having linear transition states are exemplified by studies on a biogenetic type synthesis of penicillin and, more generally, on the endocyclic attack of a nucleophile attached by a linking chain to an enone or polarized double bond.

The formation of five-membered rings is considered. In base, enones such as 4-hydroxy- and 4-amino-2-methylenebutanoates close exocyclically to give the corresponding five-membered lactones and lactams, respectively. Hydroxymethyl vinyl ketones, however, do not cyclize in base. In strong acid, hydroxy-enones close endocyclically to five-membered rings, as in the formation of cyclic ethylene ketals from acetone and ethanediol. The endocyclic closure of nucleophilic centres to polarized double bonds to give five-membered rings is impeded by restrictions of geometry; these restrictions can be overcome in acid through the generation of oxonium or protonated forms.

The rules that I have enunciated for ring closure which some have called Baldwin's rules may also be Baldwin's folly (Baldwin 1976*a, b*). However, the ease of formation of cyclic structures has some regularity or pattern which perhaps many people have known as part of the art of organic synthesis but have not formulated on paper. Before I did, Professor Eschenmoser recognized that the endocyclic transfer of groups, which was originally easily 'explained' by drawing Robinson arrows, did not proceed because of the restrictions placed on these reactions by restraints of an S_N2 displacement being a linear transition state (Tenud *et al.* 1970).

I became aware of the possibilities of extensions of that as a result of trying to synthesize penicillin. We had tried to do this biogenetically, by a route suggested independently by Arnstein in 1958 (see Abraham 1974) and also by Professor Birch (see, e.g., Birch & Smith 1958), in which the so-called Arnstein tripeptide (1) was converted into an intermediate peptide (L-α-aminoadipyl-L-cysteinyl-D-valine) (see Scheme 1) which was sequentially converted into the thioaldehyde (2). This (2) can reasonably be considered to ring close (even

though that might not be thermodynamically favourable) to the azetidinone (3) which then undergoes a simple Michael-type ring closure (or 1,4-addition) to give the thiazolidine portion of penicillin.

We decided to make the protected intermediate (2) to see whether it would cyclize to penicillin. In the light of what we found, this seemed rather a naive hope. We generated protected derivatives of the esters of (2) in two different ways but found that they do not give any penicillin as far as we can determine (and, of course, one is looking for a substance that is probably present in very small amounts, if at all). So we wondered whether the problem in this scheme arose in the step (2)→(3) or in the cyclization of (3). The azetidinone might be formed but reacting otherwise. It turned out that in our conditions the major product from (3) is, not unexpectedly, the disulphide of the corresponding ene-thiol. Therefore, we decided to degrade penicillin into the azetidinone (3). We obtained it from the phthaloylpenicillin methyl ester by a series of known reactions as a crystalline compound. The key step was deprotection of the thiol; the formation of *seco*-β-lactams is well established. This compound did not undergo the 1,4-addition.

Wolfe *et al.* (1969) had claimed to have generated a substance resembling (3) by the opening of anhydropenicillin. They made the phthaloyl-thiazolidine (5) which they treated in mild basic conditions (pH 7.4 in Me_2SO). They suggested that the thiolactone was hydrolysed (giving the corresponding acid – as the thiolate – of [3]). They then isolated the corresponding penicillin. We have been unable to repeat this work. One danger in their method is the use of a bio-

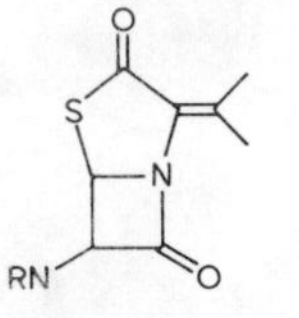

(5) R = phthaloyl

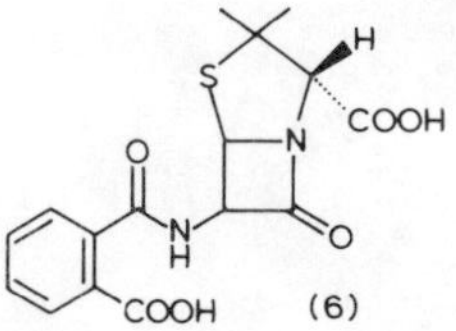

(6)

assay. Phthaloylpenicillins have low activity in the bioassay and they were developing something which bioassays as a penicillin. However, if (5) was contaminated with the corresponding phthaloylpenicillin, which has low antibacterial activity, they may have isolated the phthalamic acid (6) by hydrolysis of the imide bond (which does happen in those conditions) with a resulting increase of activity in the bioassay. We isolated the crystalline substance (3) but could not close the ring. This seems to us more substantial evidence than the suggestion of its intermediacy in the conversion of anhydropenicillin into penicillin. Why does the azetidine-thiol (3) not cyclize?

At that stage of our work, Professor Dunitz came to us from Zurich and described his work on the X-ray structure of certain amino-ketones. He suggested that when such a ketone is attacked by a nucleophile, the latter approaches at an angle of 109° to the plane enclosing the ketone (i.e. about the tetrahedral angle). When a double bond is αβ- to the ketone, similar attack by the nucleophile at an angle of 109° to the plane of the enone might explain these problems; that is, if the substituent attached to the enone system is a nucleophile, the linking chain of atoms would inhibit the approach of this nucleophile into the area of space necessary for that type of addition. This led us to wonder whether a general trend underlaid these so-called endocyclic closures. Professor Eschenmoser's work (Tenud et al. 1970) bears on this; they observed that nucleophiles in an onium salt structure do not undergo what superficially looks a reasonable reaction (Scheme 1) because of the requirement

SCHEME 1

of colinearity for the transition state in the S_N2 displacement. The extension of that was the idea that the endocyclic attack on an enone or polarized double bond by a nucleophile in a linking chain, as in (7), would depend, if there were a precise arrangement in space such as is suggested by Dunitz, on the length of the linking chain (i.e. its ability to adopt the correct spatial position). If the structures were modified, say by rotation around the =C–CO bond (see e.g. [8]), the approach of the nucleophile would occur freely into the appropriate space

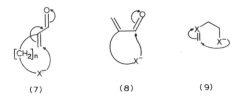

(7) (8) (9)

with respect to the ketone. Therefore, the so-called exocyclic closures would be favoured for almost any ring size (as judged from models) whereas the endocyclic closure would only succeed when the linking chain was long enough to provide for the approach in the correct position (i.e. $n = 3$ or 4 in [7]) and be difficult, or fail, if n was 2 or less. (Many examples of successful exocyclic Michael reactions are known.)

We were interested in the case of olefins (9) which contain a polarizing function. Superficially, the cyclization seems eminently reasonable. We designed several simple experiments to find out whether it was such a good method of making five-membered rings. In the endocylic mode of trigonal additions we studied, for example, base treatment of the αβ-unsaturated ester (10); the alkoxide anion so formed could exocyclically attack the carbonyl carbon atom or endocyclically attack the $CH_2=$ position (broken arrow; Baldwin *et al. 1976*).

Alcohols like (10) and compounds in which the OH is replaced by an amine group or a carbanion always rearrange exocyclically on such treatment to the lactone type of compound (11); for instance, the amino-esters (12) close to the cyclic lactam (13) by *trans*-acylation, whereas in bimolecular reactions a primary amine always adds in a 1,4-mode on the terminal olefinic carbon atom.

Apparent 5-endocyclic carbanion additions have been reported. Marino & Mesbergen (1974) claimed that strong base deprotonated the cyclohexanone (14) and gave the anion which added to acrylic ester to form the anion (15). This then cyclized to the bicyclo[4,3,0]nonane system (16). This appears to be a 5-endocyclic addition to a polarized double bond. Unfortunately the X-ray structure proved that (16) is an incorrect formulation of the product, which turns out to be (17), formed as a result of deprotonation of the vinyl carbon atom (which is often acidic in vinyl sulphides), addition and exocyclic closure.

Two other cases which we looked at in greater detail were the base-catalysed closure of compounds of the hydroxy-enone type (18) (which we could prepare easily) to give furanone (19), presumably by Michael addition to the enolate. We prepared such furanones and the corresponding hydroxy-ketones to try to do this closure but with base we have been unsuccessful. Also, base does not cause the enolate to revert to the hydroxy-ketone. The two 4-H atoms in the furanone exchange quickly with ^{2}H (from MeO^{2}H/MeO$^-$) to give (20) with no formation of (18). This lack of closure is not the result of thermodynamic considerations, which always have to be taken into account when a reaction does not work, but presumably because there is a large activation barrier to the closure of the hydroxy-enone. This may not be a good example because alkoxides are not good nucleophiles to such enones but we can get round that objection by replacing, for example, the methoxide (or t-butoxide) with the deuteriated solvent whereupon we recovered the starting enone with ^{2}H in the 4-position (21), which can only be formed by addition of alkoxide to form the enolate of the β-alkoxy-ketone which is deuteriated and reverts to the equilibrium position, namely (21). This experiment proves that (18) is amenable to an intermolecular alkoxide addition but not to an intramolecular one which would be the 5-endo-type.

Let me give one more example of nucleophilic addition to illustrate the utility of at least thinking in these terms, namely treatment of *trans*-cinnamoyl chloride (22) with hydrazine (Baldwin *et al.* 1976). In suitably mild conditions, the hydrazide (23) is formed. However, our attempts to close this to the pyr-

azolone (24) by heating the free base and its sodium salt to high temperature failed; the hydrazide decomposed. Treatment of the acid chloride with methanol to form the ester and subsequent treatment of that with hydrazine at 50 °C rapidly gives the pyrazolone, the reason being that the acid chloride has been so deactivated that the hydrazine does not add to the ester but adds by a Michael reaction β to the ester to give the hydrazino-ester (25), which undergoes exocyclic closure. Heterocyclic chemists have long known that in making such compounds by addition of a two-atom unit to an enone it is best to form β-bond first.

In strong acid, the enone ring cyclizes. This acid-catalysed closure is probably best illustrated by the formation of cyclic ethylene acetals. Ketones readily form an acetal with ethanediol in acid. The usually accepted mechanism is formation of a hemiacetal (26) which loses water in acid to give the oxonium system (27) which then closes (Scheme 2)—as written, this is an endocyclic 5-membered closure. The same applies to the protonation of the hydroxy-enones

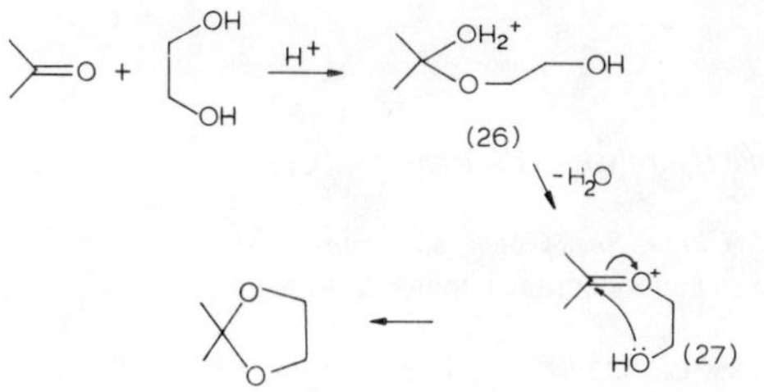

SCHEME 2

(18) to give vinylogues of (27). Why should base catalysis cause such a problem for ring closure? There are other mechanisms for formation of five-membered cyclic acetals, such as direct displacement by an exocyclic tetrahedral 5-displacement, but I do not like that mechanism. Another is deprotonation of the oxonium ion (27) to the *enol*-ether which undergoes what is formally an exocyclic addition (Scheme 3), but presumably via the same intermediate as (27). We can

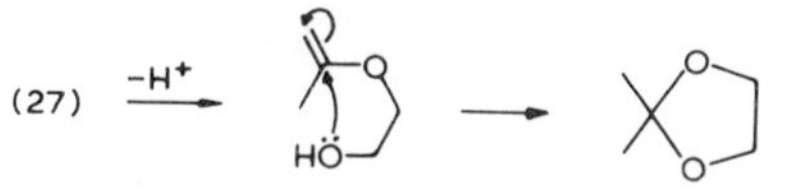

SCHEME 3

eliminate that mechanism for the formation of cyclic acetals since we have observed no exchange α to the ketone during acetalization. The explanation I can offer for this that such oxonium species as (27) have different geometrical requirements for the alignment of the nucleophile, the hydroxy group, to the double bond in the addition. They may be approached from the direction perpendicular to the plane of the oxonium ion (i.e. they are treated like carbonium ions). There are two restrictions: on the Dunitz approach, the OH group tends to go further back than it would need; secondly, rotation around the Me$_2$C–O bond can release the geometrical restraint placed on the endocyclic addition. N.m.r. spectra of an *O*-methylated ketone (oxonium ion) suggest that the barrier to these rotations is low, about 20 kcal/mol, but these experiments do not seem to differentiate between rotation and a simple trigonal inversion mechanism (which does not involve rotation around a formal double bond). The same comments apply to the corresponding immonium ions which close readily without an enamine intermediate into oxazolidines.

The ideas about the problem of endocyclic closure seem to be borne out in these simple systems; they show that there are geometrical problems in closing five-membered rings by those means when nucleophilic centres are being added to polarized double bonds. Apparently, these problems can be overcome in the oxonium or protonated forms and the rings can be closed.

References

ABRAHAM, E. P. (1974) *Biosynthesis and Enzymic Hydrolysis of Penicillins and Cephalosporins*, University of Tokyo Press, Tokyo

BALDWIN, J. (1976a) Rules for ring closure. *J. Chem. Soc. Chem. Commun.*, 734–736

BALDWIN, J. (1976b) Approach vector analysis: a stereochemical approach to reactivity. *J. Chem. Soc. Chem. Commun.*, 738–739

BALDWIN, J., CUTTING, J., DUPONT, W., KRUSE, L., SILBERMAN, L. & THOMAS, R. C. (1976) 5-Endo-trigonal reactions: a disfavoured ring closure. *J. Chem. Soc. Chem. Commun.*, 736–738

BIRCH, A.J. & SMITH, H. (1958) Oxidative formation of biologically active compounds from peptides, in *Amino Acids and Peptides with Antimetabolic Activity (Ciba Found. Symp.)* (Wolstenholme, G. E. W. & O'Connor, C. M., eds.), pp. 247–260, Churchill, London

MARINO, J. P. & MESBERGEN, W. B. (1974) Cycloadditions of allyl anions. I.A regiospecific and stereospecific pentannelation reaction. *J. Am. Chem. Soc. 96*, 4050–4052

TENUD, L., FAROOQ, S., SEIBL, J. & ESCHENMOSER, A. (1970) Endocyclische S_N-Reaktionen am gesättigten Kohlenstoff. *Helv. Chim. Acta 53*, 2059

WOLFE, S., LEE, W. S. & FERRARI, C. (1969) Mercuric acetate oxidation of an anhydropenicillin. Anhydro-α-phenoxyethylpenicillene, a novel antibacterial agent. *Tetrahedron Lett. 39*, 3385–3388

Discussion

Woodward: Your point about the enone (18) being amenable to intermolecular alkoxide addition but not to intramolecular addition of the 5-endo-type seems dubious to me. Is there anything bad about an allenic enol (e.g. [28])?

Baldwin: Such compounds have been made in strongly basic conditions.

Woodward: But instances are known in which such enolates are produced in relatively mild basic conditions.

Baldwin: If so, that destroys the argument that this demonstrates the ease of 1,4-addition. However, another experiment in this series also supports this. When we made a system in which the alcohol (as an ester) is in such a position as to make the closure to a five-membered ring exocyclic (e.g. [29]), hydrolysis in base gave only the corresponding cyclic ether (30). The electronic arrangements in this reaction are probably similar to those in the formation of the furanones (19), but (29) does not contain an α-oxy substituent. Though we could not close compounds like (18) in an endocyclic mode, they spontaneously close exocyclically from the alkoxide.

Baker: How do Rosenblum's conclusions (1974) on the ready reaction of the

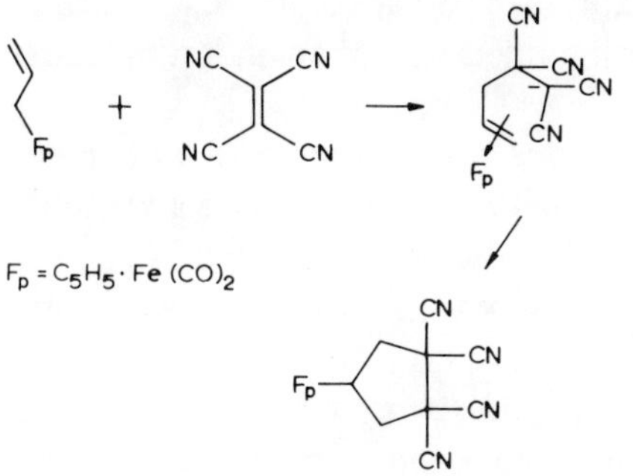

SCHEME 1 (Baker)

allyl–iron complex with tetracyanoethylene to give a cyclopentane (Scheme 1) compare with these ideas?

Baldwin: According to our ideas, the formation of the cyclopentane should be unfavourable, if that is the mechanism. I should add that these considerations do not apply to those reactions that can be written formally as having biradical character because, in principle, one can release the restraints — the lack of rotation of the double bond and perhaps the requirement of a transition state for the addition to the double bond— in such biradical reactions.

Brown: If one considers the reverse process of ring opening in terms of stereoelectronics, one can predict for, say, penicillin deprotonated at the α-carboxy carbon atom that the conjugated anion and the bond to be broken are orthogonal and overlap is poor. One can derive this result and also make a prediction about Rosenblum's cyclization and dioxolan synthesis without considering reaction trajectories. In both these last two examples overlap of the reaction components is possible.

Baldwin: Why?

Brown: Let us consider a five-membered ring. For the bond βγ to carbonyl to break, charge must be transferred from an orbital on the α-carbon atom which is orthogonal to it. This not so for ketals.

Baldwin: Why should the lone-pair on oxygen be *anti*-periplanar to that molecule?

Brown: It doesn't have to be *anti*-periplanar but it is not orthogonal. In stereoelectronic terms it is much easier to consider the product and follow the reverse course.

Breslow: Yes, providing it gives the same result. The new π-bond being made is not the π-bond that will be found in the product. That is a serious difficulty.

Brown: If stereoelectronics are important, then perhaps endocyclic routes to and from five-membered rings where the intermediate carbanion is tetrahedral (e.g. CF_3-stabilized) are possible.

Baldwin: Why? In a five-membered ring one is concerned with a fragmentation which is induced by some carbonyl-stabilized anion. There is always the problem of the alignment of this 'orbital' with the breaking bond.

Brown: Are your hypotheses going to work best in those cases where the reaction is not particularly exothermic?

Baldwin: Most of the problems that we synthetic chemists encounter in building a complicated molecule with many functional groups relate to the competition of reactivity between those groups. I have tried to explain that there are several ways of forming ring structures with polyfunctionality. The idea that exocyclic closures of trigonal systems will in general be much faster than the corresponding endocyclic closures is a good general principle (with some exceptions). I am not saying that one cannot form or open a five-membered ring by what is formally an endocyclic process. The reason a ring does not close is an activation barrier. If one can design an exothermic process, one can surmount that barrier.

Golding: In at least one instance nature may have anticipated the 'Baldwin rules'. In the biosynthesis of porphobilinogen δ-aminolaevulinate dehydratase might have aligned two molecules of δ-aminolaevulinic acid (δ-ALA) and attempted production of porphobilinogen as in Scheme 1. However, the cycliza-

SCHEME 1 (Golding)

tion step is likely to be stereoelectronically impeded. The enzyme actually condenses one δ-ALA with the ε-NH_2 group of a lysine, then effects an intermolecular condensation with the other δ-ALA and finally an unconstrained cyclization occurs (Scheme 2) (for a detailed discussion see Cheh & Neilands 1976).

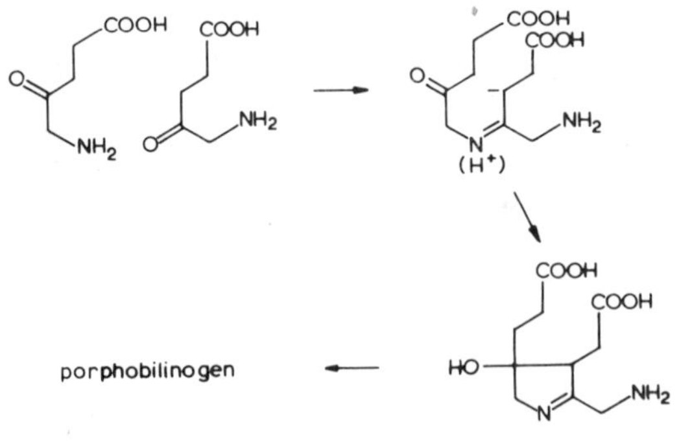

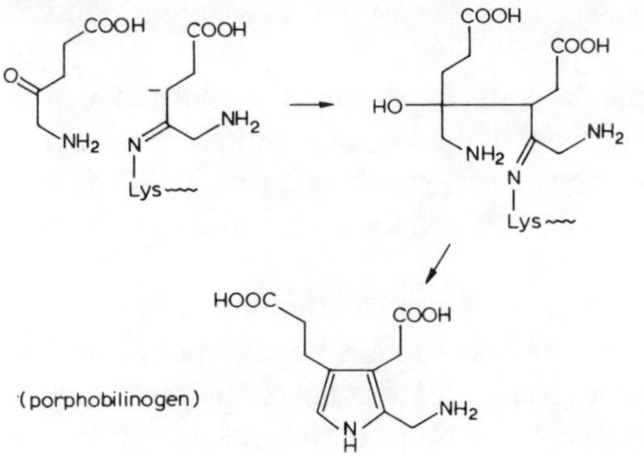

RULES FOR RING CLOSURE

SCHEME 2 (Golding)

Baldwin: But it is exocyclic to the carbonyl. This is essentially the same problem, because the electrons are concentrated in the plane perpendicular to three atoms of the enamine.

Golding: How readily do enamines such as (31) cyclize? The cyclization in Scheme 3 is analogous to that in my Scheme 1 and is predicted to be unfavourable.

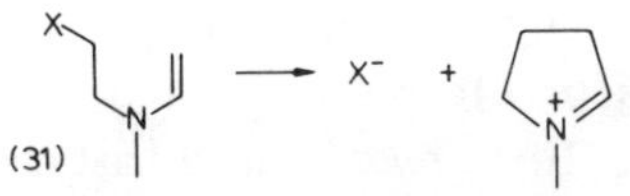

SCHEME 3 (Golding)

Baldwin: It should not be possible to make a cyclopentanone from the enolate of the ketone (32) if X were a leaving group, or at least that should not be

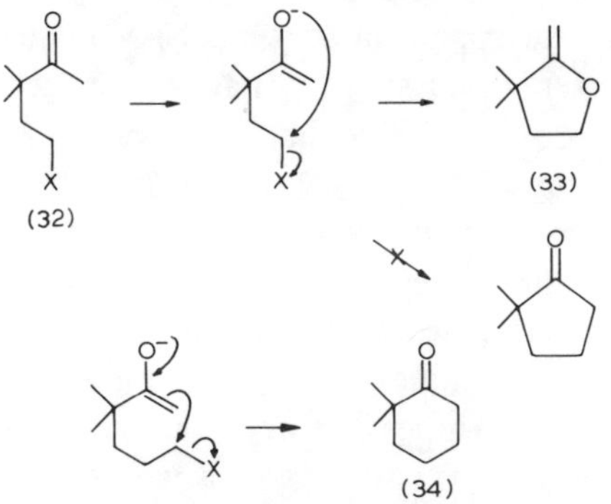

a favourable process relative to the alternative which is the closure of the enolate ion. When X is Br, treatment with various bases and metal ions gives in a clean reaction only enol-ether (33). But if you put in another carbon atom, cyclohexanone (34) is the only product.

Golding: The carbanion in my Scheme 1 is constrained because it wants to interact with the C=N bond.

Baldwin: I should point out that by formally writting the enolate anion as such I am not implying any involvement of the electrons which are part of the enolate resonance, because there are two sets which are orthogonal, those of the enolate and the two lone-pairs on oxygen, and the latter suffer no restriction with respect to ring closure.

Baker: The work of Sergeant & McLaughlin (1970) forms a measure of the effectiveness of this type of closure, in terms of double-bond participation. The distance of the double bond from the reaction centre is not crucial but the alignment of the double bond with that centre is. Can we explain Rosenblum's results by saying that although the alignment may not be completely favourable the reaction takes place because of other favourable factors, e.g. intramolecular nucleophilic attack on a coordinated double bond?

Baldwin: The use of iron and tetracyanoethylene raises questions about the mechanism for that reaction. The alignment of the reacting atoms in an attack on a double bond is absolutely critical to the success of the reaction. The case of the enolate shows that with enolates or any isoelectronic analogue such as amide anions or anilines one has to consider formally two sets of electrons which are mutually orthogonal: one set perpendicular to the plane and the other (non-bonding electrons) usually in the plane. In many cases, at least in intramolecular ring closures, it is a question of which of these is best aligned. (These are always much better than the corresponding set.)

Woodward: With regard to what you said about acetalization, some precision in language is needed. You mentioned the possibility of a problem in the base-catalysed endocyclic closure when a nucleophile attacks a polarized double bond. But in just this case that you were discussing you have a polarized double bond and a nucleophilic centre. Would you agree that the more nearly the cyclization that you are considering resembles an S_N2 displacement the more the proposition you made will be true? In this whole area we shall in future be faced with the situation that we have been thinking about now for 30 years with regard to the whole question of displacements at a satured carbon atom, namely that there is whole range of possibilities, from S_N2 to pure S_N1, a range which also exists for these cyclizations. At the extreme, when the reaction resembles an S_N2 displacement—as will be the case when one works with anions—one would expect precisely what you said: cyclization will be extremely difficult. This difficulty will diminish as the reaction becomes more like an S_N1 reaction. Depending on the substituents and many other factors, one can never be sure what case one is dealing with.

Baldwin: It is always easy to say that one cannot achieve a certain reaction because of some formally stated rule. You are saying that for those reactions that are like S_N2 processes the departure of the leaving groups in the S_N2 is strongly coupled with the incoming group. But, for example, for the allylic cation formed by loss of the halide from allylic halide and closure, the geometric restraint is removed.

Woodward: Reaction of an anion with a double bond may be regarded as close to an S_N2 displacement—one of the two bonds of that double bond being displaced, by backside attack, with the stereochemistry that you say is hard to achieve in certain cases. That gives a perfectly rational basis for the difficulty of such reactions. But if the centre that is being attacked in the cyclization is a pure carbonium ion, cyclization will present no difficulty.

Baldwin: That fits in with the ability of these five-membered rings to close endocyclically in acid but not in base.

Ramage: That means that the trajectory of approach onto the π-system will vary between 90 and 109° according to the type of reaction.

Barton: In the acid-catalysed reactions you don't know whether there is a β-hydroxyketone intermediate in low concentration which is then displaced.

Baldwin: That is possible.

Baker: Why should it displace?

Barton: In the benzylic case, it is protonated to give an incipient stabilized cation. This is Professor Woodward's point.

Woodward: It seems to me that the height of the barrier to the cyclizations will depend in each case on the precise mechanism for reaction. That may often be difficult to know.

Prelog: And the mechanism may differ between analogous cases. With respect to steric requirements postulated by Bürgi & Dunitz which prevent the reaction when five-membered rings should be formed, you said that many examples were known for six-membered ring formations in which these restrictions did not apply. Are there entirely homologous cases in which six-membered but not five-membered rings are formed?

Baldwin: The amine (35) preferentially closes to the piperidine and the anion (36) closes by Michael addition to give the six-membered product. The phenol (37) closes easily to the cyclic ether (38) in base.

Ramage: In a cyclization such as that of the amino-acrylic ester (35), it is extremely important to consider the thermodynamics of the process. For example, in lysergic acid chemistry treatment of the corresponding piperidine ring with acetic anhydride opens the ring and leads to subsequent cyclization (by the reverse mechanism) to the lactam; in this case thermodynamics control the reaction, not only the trajectory of the approaching nucleophile.

Baldwin: True; any statement about the transition state is a statement about

(35) (36) (37) (38)

the kinetics of the reaction. In the few cases we have studied so far, we have tried to establish that the thermodynamics do not control the path.

Breslow: Another way of putting your argument, Professor Baldwin, is to say that the stereochemistry of the normal transition state has some geometrical requirements which are not easy to meet. However, it may be that the stereochemistry of the normal transition state does not represent such a problem and that, at earlier stages, the reactants cannot follow the low-energy pathway; as the approach would then have a higher energy barrier the transition state would be in a different place—in other words, the normal transition state cannot be reached. Is the trajectory a serious problem at an earlier stage or is the normal transition state strained by being constrained in a ring?

Baldwin: But are these separate things?

Breslow: They can be; there can be a new transition state earlier in the reaction if the mechanism changes in the sense that the reactants cannot follow the normal pathway.

Baldwin: In one case, such as the base-catalysed closure of the hydroxy-enones, the deuteriation results suggested that an external nucleophile, an alkoxide, could approach because of its freedom of motion and attack the enone. If we accept what you say as a working hypothesis, the internal alkoxide cannot even get onto the early path that leads to the transition state.

Breslow: According to one approach, we can consider the standard transition state and ask how much it is strained. Alternatively, the trajectory exerts its influence earlier in the reaction so that the reactants cannot get onto the low-energy trajectory to the normal transition state.

Baldwin: I had not thought of it in that way. I was more concerned with geometry than thermodynamics. In Dunitz's early picture when the alkoxide could not approach the double bond in its particular trajectory those reactions would not go. I had not considered the position of the transition state with respect to the total reaction coordinate. The inhibited basic closure may have late transition states that look very much like the products and in protonating the enones we are shifting the transition state close to the starting material.

Cornforth: Some years ago Bonthrone & I (1969) showed that to get high yields in the methylenation of catechols by dichloromethane (Scheme 1) one had to arrange conditions so that the steady-state concentration of the chloro-

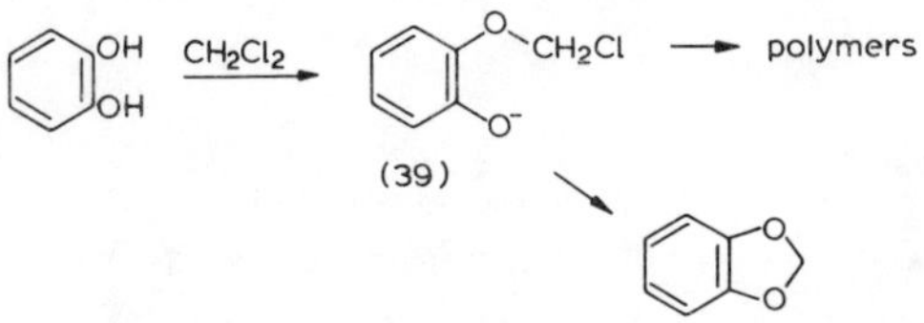

Scheme 1 (Cornforth)

methoxy intermediate (39) remained low. Otherwise linear macrocyclic polymers are formed. The best reason I could deduce for the strange preference for an intermolecular condensation was that the oxygen atom is further away from the carbon atom than is normal in a tetrahydrofuran ring closure.

References

Bonthrone, W. & Cornforth, J. W. (1969) The methylenation of catechols. *J. Chem. Soc.*, 1202–1204

Cheh, A. M. & Neilands, J. B. (1976) The δ-aminolevulinate dehydratases: molecular and environmental properties. *Struct. Bond. 29,* 123–169

Rosenblum, M. (1974) Organoiron complexes as potential reagents in organic synthesis. *Acc. Chem. Res. 7,* 122–129

Sergeant, G. D. & McLaughlin, T. E. (1970) Nucleophilicity of remote carbon–carbon double bonds. The significance of transition-state colinearity. *Tetrahedron Lett.,* 4359–4361

General discussion I

PEPTIDE SYNTHESIS

Kenner: An aspect of synthesis which excited Sir Robert when it was first published (Merrifield 1963) was solid-phase synthesis of peptides. After a good deal of controversy, the consensus now is that the method is useful for the synthesis of peptides with medium molecular weight. Various improvements have been suggested; the most important may be Sheppard's replacement of Merrifield's polystyrene matrix by a cross-linked polyacryldimethylamide matrix (Atherton *et al.* 1975). In Sheppard's view some of the defects in the conventional Merrifield synthesis are due to the incompatibility of the non-polar matrix and the polar solvents most suitable for elongation of the polypeptide chain. Gait & Sheppard (1976) have extended the application of this more polar polymeric support to synthesis of oligonucleotides. The use of these polymeric supports allows a certain measure of automation of some of the more routine manipulations. Are they likely to have wide use in organic synthesis?

Barton: I always feel a bit depressed after hearing you talk about the synthesis of polypeptides—so much elegant and painstaking work goes into the formation of just one bond which, from an organic chemist's point of view, is rather a dull bond. The question is how can one make the amide bond more quickly and in 100% yield so that one can get on with something more interesting! For that reason I advocate automation of the process.

Woodward: You would be much more interested if there were a better method of making amide bonds, in good yields, every time, without racemization.

Todd: No method that I know of, whether solid-state or not, synthesizes large molecules such as polynucleotides and proteins with any degree of precision when the molecular weight is about 10^6. These methods may facilitate matters not by making the natural compounds but by throwing light on the action of enzymes and proteins on smaller molecules.

Solid-state methods will not be valuable until we can solve that problem; then we may be able to make synthetic enzymes.

Chain: Why should one want to synthesize enzymes when one can get them readily from natural sources?

Todd: Enzymes won't catalyse every chemical reaction.

Chain: No; neither will synthetic ones. Another point is that if one is considering the synthesis of peptides for pharmaceutical purposes, unfortunately few peptides can be used pharmaceutically because most peptides are destroyed as soon as they are injected into the body. The interesting peptides from the hypothalamus, for instance, never reach the point where they should act when they are administered. This approach to peptide synthesis raises great difficulties.

Kenner: Even if the value of synthetic peptides as drugs is limited, they are valuable for understanding hormone action and the interaction between small and large molecules *in vitro*.

Barton: But an economic synthesis of insulin would be of great value.

Kenner: The CIBA-GEIGY group (Sieber *et al.* 1974, 1977) achieved a superb synthesis of human insulin, but this is not an economic proposition at present. There is, however, the possibility of making active peptides smaller than insulin. Returning to Sir Ernst's point, I believe that a future application of peptide work can be seen in the enkephalin–morphine story. In that case the non-peptide, morphine, was known long before the natural peptide which binds to the opiate receptor. In general one can imagine devising stable non-peptide compounds, once a receptor's requirements have been defined by study of synthetic peptides.

Chain: However, it is the larger peptides from lipotropin which are more active than the enkephalins.

Woodward: How far can the automated methods at present be used in a practical way?

Kenner: In general they can, with care, be used to make peptides consisting of 15–20 residues. An impressive synthesis of corticotropin (ACTH) has been reported (Yamashiro & Li 1973).

Woodward: How much material can one prepare?

Kenner: The process is more suited to rapid preparations on the microscale, but the Armour Laboratories have described a synthesis of ACTH on the 100 g scale (Colescott *et al.* 1975).

Prelog: Sir Robert had a high opinion of Merrifield's methods for preparing polypeptides and thought that nature works in a similar way. I am somewhat sceptical about that idea, because in an enzymic synthesis the substrate is not bound to the outside of a macromolecule. Instead, many weak bonds are strategically distributed in a cavity of the enzyme protein, which thus not only catalyses the reaction but also controls the stereospecificity through the chirality of the cavity. The wrong enantiomer of the substrate is either rejected or converted into the correct one. In contrast, the reactants in Merrifield's syntheses are on the 'outside' of the solid phase. The final solution of the problem of peptide synthesis will probably be some sort of simulation of natural synthesis.

Eschenmoser: Nature is always quoted as the example in matters of specificity in synthesis. Is this assertion that nature synthesizes proteins with nearly perfect specificity a fact or an extrapolation? We find and isolate natural proteins mainly on the basis of their biological activity; but in the isolation procedures, we might be throwing away proteins that have no function. Does the apparatus of protein synthesis make non-functional proteins?

Prelog: Surely not; nature cannot afford to have too many useless proteins in the cell. Practically all proteins that are present probably have a function even if that function is at present not evident.

Woodward: Is that a statement of fact or belief?

Prelog: Unfortunately this is only my belief! But if one considers the many activities controlled by proteins and the number of proteins the cell badly needs, one *feels* that little space remains for useless proteins. They would also be eliminated by evolution.

Golding: According to one theory of insect resistance to DDT the dehydrochlorinase which removes HCl to give inactive DDE was a useless protein which found a use when DDT appeared.

Todd: If present work establishes what is going on when something is taken up and adsorbed on an enzyme and the factors that control those steps are understood, then we may be able to make artificial enzymes. The rest of the protein molecule in the natural enzyme cannot be ignored but the rest of the protein molecule of an enzyme need not be what is there naturally or anything like so precise in its structure if one is proposing to mimic, so to speak, *in vitro* reactions brought about at the enzyme's active site. Therefore I foresee a future for the synthesis of relatively small peptides that incorporate much of the 'working end' coupled to a comparatively easily-made synthetic protein which need not have the precise constitution of the enzyme. These solid-state methods may come in here. This is where I differ from Sir Ernst. Many chemical reactions are basically similar and probably do not need the specificity of a natural enzyme. These methods could be used to make artificial enzymes with which one might catalyse a variety of reactions.

Chain: But for reactions more complicated than hydrolysis you are talking not of single enzymes but of multienzyme complexes in which carbohydrates, lipids and so on are important; absence of just one of these components may suppress the activity.

Ramage: With regard to the difference between solid-phase and solution-phase peptide chemistry, even the best methods at present, namely those of Sheppard *et al.* with their new resin, do not give 100% incorporation of amino acids into the growing chain. Most importantly, the incorporation seems to depend on the particular residue being added at a particular time: there are good sequences and bad sequences. We need to distinguish between these. A good method of making the amide bond would make the choice between solid-phase and solution almost irrelevant. In *Perspectives in Organic Chemistry* (1956) Professor Woodward alluded to an approach to the synthesis of peptides that avoids solution methods.

Kenner: In protein synthesis we need an efficient method of joining sizeable pieces with molecular weights of 5000 or more. In the nucleotide field suitable enzymes are available, but this is not so for polypeptides and there is nothing comparable to the specific association of nucleotides through base-pairing. In principle, joining the components needs to be turned into an intramolecular reaction. Although making an amide bond is apparently a trivial reaction, over 130 methods have been described in the peptide field (Stelzel & Wendlberger 1974) and yet there is still scope for Sir Derek to provide the good method. The reaction is not so easy when the carboxy group may be folded in part of one molecule and the amino group in the other, and throughout the system there are indole rings, S atoms and the like with intrinsically lower nucleophilicity than the amino group but perhaps great accessibility. The solution to the problem may lie in joining the compounds by a weak link, such as a disulphide, to bring together two activated functions in a favourable environment for intramolecular reaction.

Woodward: You have defined an extremely important problem, to which one could think several general methods of approach; yet there might always be the lurking suspicion that each case is special.

Barton: Aren't all those 130 methods based on the amine function or its equivalent being the nucleophile and the carboxy function or its equivalent being the electrophile?

Kenner: Most are clearly in that category, but there is a group of methods based on the reaction of tervalent phosphorus with the amino group followed by the addition of the carboxy group (e.g. reaction 1)

$$(EtO)_2P \cdot NHR^1 + R^2COOH \rightarrow R^2CO \cdot NHR^1 + HPO(OEt)_2 \qquad (1)$$

and there is also conversion of the amine into an isocyanate followed by addition of the carboxy component and expulsion of CO_2 (Goldschmidt & Wick 1952).

Barton: Isn't the N atom then being nucleophilic?

Kenner: Yes, that is essentially true even in these cases. It would be difficult to reverse the polarity to make the amino group the electrophile.

Baldwin: If current methods are not going to be successful in making large polypeptides (such as insulin) cheaply, could not the isolated ribosome or its components be used to process the gene *in vitro?*

Kenner: K. Murray in Edinburgh is exploring the synthesis of insulin precursors in collaboration with ICI Corporate Laboratory.

Chain: Berg (1977) has reviewed genetic engineering but his article exclusively contains promises: for instance, that genetic engineering will soon lead to discoveries of great practical importance, by which Berg presumably refers to the possibility of breeding new plants with an industrially-interesting composition (containing more oil or protein) or to the production of insulin and other mammalian peptide hormones by bacterial fermentations after transfer of the genes coding for these proteins from mammalian to bacterial genomes; many of his colleagues in the area of genetic engineering have made such completely speculative remarks. He presented no trace of evidence that these things can be done—the whole article is just a request for money to do such research. I mention this because over the past 30 years this has been the trend of scientific approach, to which I take great exception. Let them do the work and see what they produce.

Battersby: I want to add support to what Lord Todd and Professor Kenner said about the possibility of synthesizing molecules smaller than proteins which will do some of the jobs that enzymes do. I am thinking of mini-enzymes in the molecular weight range of about 1000 to 3000, with an 'active site' and a cavity to hold the substrate. These molecules need not be polypeptides; indeed, the molecules explored so far or currently being studied usually do not contain amino acids. Being realistic, one has to accept that work in this area will take a lot of time and we shall be fortunate initially to reproduce *one* feature of enzymic action. So I am sure Professor Cornforth will be very happy if he is able to hydrate a double bond in an anti-Markownikoff way and others equally so to be able to work with bound oxygen. These are just examples from many such studies.

Sondheimer: This discussion is showing up something which is perhaps obvious, namely that organic synthesis encompasses an enormous range of different activities. Luckily, not all synthetic organic chemists believe that a certain activity is the most important one — otherwise we would all be doing the same thing! Whether one makes polypeptides, whether one tries to copy a certain complicated enzyme which exists in nature, whether one should try to make a simpler compound which has some of the activities of an enzyme, whether instead of trying to make an amide bond one tries to develop new synthetic methods, each of these activities is of great interest. One cannot give priority to any one in particular.

References

ATHERTON, E., CLIVE, D.L.J. & SHEPPARD, R. C. (1975) Polyamide supports for polypeptide synthesis. *J. Am. Chem. Soc. 97*, 6584-6585

BERG, P. (1977) Recombinant DNA research can be safe. *TIBS (Trends Biochem. Sci.) 2*, N25–N27

COLESCOTT, R. L., BOSSINGER, C. D., COOK, P. I., DAILEY, J. P., ENKOGI, T., FLANIGAN, E., GEEVER, J. E., GROGINSKY, C. M., KAISER, E., LAKEN, B., MASON, W. A., OLSEN, D. B., REYNOLDS, H. C. & SKIBBE, M. O. (1975) Large scale solid phase synthesis of ACTH, in *Proceedings of Fourth American Peptide Symposium* (Walter, R. & Meienhofer, J., eds.), pp. 463–467, Ann Arbor Science Publishers, Ann Arbor, Michigan

GAIT, M. J. & SHEPPARD, R. C. (1976) A polyamide support for oligonucleotide synthesis. *J. Am. Chem. Soc. 98*, 8514–8516

GOLDSCHMIDT, S. & WICK, M. (1952) Über Peptid-Synthesen, 1. *Liebigs Ann. Chem. 575*, 217–231

MERRIFIELD, R. B. (1963). *J. Am. Chem. Soc. 85*, 2149–2154

SIEBER, P., KAMBER, B., HARTMANN, A., JÖHL, A., RINIKER, B. & RITTEL, W. (1974) Totalsynthese von Humaninsulin unter gezielter Bildung der Disulfidbindungen. *Helv. Chim. Acta 57*, 2617–2621

SIEBER, P., KAMBER, B. A., JÖHL, A., HARTMANN, A., RINIKER, B. & RITTEL, W. (1977) Totalsynthese von Humaninsulin. IV. Beschreibung der Endstufen. *Helv. Chim. Acta 60*, 27–37

STELZEL, P. & WENDLBERGER, G. (1974) in *Methoden der organischen Chemie* (Houben–Weyl). vol. 15, ch. 41, Georg Thieme, Stuttgart

YAMASHIRO, D. & LI, C. H. (1973) Adrenocorticotropins, 44. Total synthesis of the human hormone by the solid-phase method. *J. Am. Chem. Soc. 95*, 1310–1315

Some recent developments in quantum organic chemistry

MICHAEL J. S. DEWAR

Department of Chemistry, The University of Texas at Austin

Abstract The present status of attempts to calculate chemical behaviour in organic chemistry in a quantitative manner is reviewed. Results given by a new semi-empirical SCF MO method (MNDO) are reported. Specific topics discussed include the calculation of molecular vibration frequencies, entropies, specific heats, entropies of activation, kinetic isotope effects, and the mechanisms of several organic reactions, in particular the Diels–Alder reaction and the Cope rearrangement.

The modern era of theoretical organic chemistry began 50 years ago with the introduction by Robert Robinson of his electronic theory of organic chemistry. Although this seemed to be overshadowed a few years later by Pauling's resonance theory, we can now see that the latter had no real basis in wave mechanics but represented in effect no more than a translation of the older treatment into a new and less convenient terminology. Sir Robert's extraordinary insight is shown by the effective survival of his treatment, developed before the emergence of wave mechanics, for the first three decades of the wave mechanical age. It was finally displaced by a treatment based on the application of perturbation theory to the molecular orbital approximation which reached fruition in the early 1950s (see Dewar 1969; Dewar & Dougherty 1975).

Both these treatments were essentially qualitative and suffered from corresponding limitations. In recent years, however, a further step has been taken by the development of treatments that allow chemical behaviour to be predicated in a quantitative manner. Although the accuracy presently attainable is by no means as good as one would like, the results have nevertheless proved good enough to provide valuable information about the course of chemical reactions and to lead to revisions of several generally accepted ideas.

Confusion has been caused in this area by the publication of numerous papers that purport to describe calculations of this kind but which in fact are based on procedures that are unsatisfactory for the purpose. The trouble lies in the fact that only rough solutions of the Schrödinger equations can be obtained for typical organic molecules, so the calculated energies are in error by, chemically speaking, vast amounts. Such procedures can be used at best empirically, in areas where they have been tested and shown to give satisfactory results, owing to a fortuitous and unpredictable cancellation of errors. In the absence of such tests, the results of such calculations are meaningless. Unfortunately, tests have shown that the simpler procedures are nearly all far too inaccurate for chemical purposes, and the more sophisticated ones are so costly that they have not even been tested. Furthermore, it has been accepted practice to make reasonable assumptions about molecular geometries because the cost of complete geometry optimizations is so great. Calculations involving such assumptions are essentially worthless, as our own work has already demonstrated.

My group has been concerned for the last 15 years with an attempt to develop a treatment accurate enough and reliable enough to be of real predictive value in organic chemistry and also cheap enough to be applied to molecules of real chemical interest. Our first success came with a treatment (Bingham *et al.* 1975) which we termed MINDO/3, MINDO standing for modified INDO, INDO being a method first introduced by Pople (see Pople & Beveridge 1970). Recently, we have developed a still better treatment, MNDO (Dewar & Thiel 1977*a, b*) based on the NDDO approximation (see Pople & Beveridge 1970). Both these treatments have been thoroughly tested by calculations of numerous properties of hundreds of molecules and the results have been, or are being, published in full detail, so there can be no misconceptions about their potential.

The main value of theoretical calculations naturally lies in areas where experimental data are lacking. We can, for example, study the structure and properties of stable species by standard experimental procedures but we cannot follow the changes that occur during the course of a chemical reaction because the time involved is so short (about 10^{-13} s). Our theories of reaction mechanisms are therefore based on roundabout reasoning from experiment combined with ideas derived from qualitative theories of molecular structure. This is an area where an accurate quantitative theoretical treatment could clearly be of major value. However, a serious difficulty arises in attempts to study such problems, because any treatment we can use is necessarily empirical. How can we trust the results given by such a treatment in areas where it has not been tested? The answer, of course, is that we cannot be sure of the results. The most we can do is to test our method as thoroughly as we can in all connections where experimental data are available, and also do calculations for as many reactions as possible that have been studied experimentally. We should, for example, be able to reproduce the one measurable quantity related to the intermediates in a reaction, i.e. the thermodynamic properties of the transition state. Unless these tests have been done, and unless our treatment has survived them, we certainly cannot put any trust at all in conclusions drawn from calculations using it. I must emphasize again that *no* method other than ours has met these requirements. What about MINDO/3 and MNDO?

TABLE 1

Mean absolute errors in calculated heats of atomization

Class of compound	Number	Mean error (kcal/mol)	
		MINDO/3	MNDO
All compounds	209	15.7	7.5
Hydrocarbons	58	9.7	6.0
aromatic	5	12.1	1.7
acetylenes	6	13.5	6.6
cyclopropanes	8	11.7	5.2
cyclobutanes	7	8.6	18.7 (13.5^a)
cyclic	32	11.9	8.0
polycyclic	5	22.3	2.9
Nitrogen compounds (CHN)	24	17.3	6.5
cyanides	8	19.6	4.6
with N–N bonds	11	29.5	8.1
Oxygen compounds (CHO)	39	6.8	5.2
Fluorine compounds	71	24.8	9.8
All CHON compounds, excluding 4-membered rings, t-butyl, NO bonds	122	10.9	5.0

^{a} Excluding cubane.

Table 1 shows average errors in the calculated heats of formation (from gaseous atoms) of 200-odd molecules, including ions, radicals and carbenes, for MINDO/3 and MNDO. The molecules were chosen with specific emphasis on types that had proved especially 'difficult' since our object was to test the two procedures, not whitewash them. The errors for a random sample would be less than one-half those for this set. Even here, however, and even for MINDO/3, the errors are less by several orders of magnitude than those that would be given by alternative procedures applicable to molecules of this size (up to 30 atoms). Table 2 shows a similar comparison for bond lengths and angles.

As yet only a few reactions have been studied with MNDO. The average er-

TABLE 2

Mean absolute errors in calculated bond lengths and angles

Bond or angle	Number	Mean error (Å or degrees)[a]	
		MINDO/3	MNDO
All CHON bond lengths	228	0.022	0.014
C–H	56	0.019	0.005
C–C	96	0.016	0.012
C–N	17	0.029	0.010
N–N	9	0.074	0.032
O–O	3	0.117	0.043
Bonds to fluorine	55		0.040
Bonds to boron	8		0.008
All bond angles (CHON)	91	5.6	2.8
C–C–C (acyclic)	12	5.9	2.0
All angles at N	15	7.1	3.2
All angles at O	9	10.7	8.5
Angles involving F	47		3.7

[a] 1 Å = 0.1 nm.

ror in the activation energies calculated by MINDO/3 for 24 reactions of many different kinds, with activation energies ranging from near zero to over 80 kcal/mol, was about 6 kcal/mol, less than the average error in the heats of formation calculated for the reactants and products. The mechanisms predicted for all the 209 reactions we have studied have also all been consistent with the experimental evidence, though not by any means always in agreement with those given in current text books.

Other properties for which MINDO/3 and MNDO have given reasonably satisfactory results include ionization potentials, dipole moments, electric polarizabilities and hyperpolarizabilities, molecular vibration frequencies and derived thermodynamic properties (entropies, specific heats), nuclear quadrupole coupling constants, and ESCA chemical shifts (see Dewar 1975).

The results for molecular vibration frequencies are particularly interesting (Dewar & Ford 1977). MINDO/3 usually reproduces these to within 5%. C–H stretches are an exception, the calculated values being systematically too high by 500 cm^{-1}. With this correction, the errors are $<2\%$. No other available procedure is as good as this and the MINDO/3 calculations can, moreover, be done quickly and easily even for quite large molecules. MNDO gives even better results at little extra cost. Calculations of this kind may, therefore, be useful in assigning vibrational spectra and interpreting structures. A good example is beryllium borohydride, for which several structures have been suggested on the basis of electron diffraction studies. MNDO predicts (Dewar & Rzepa 1977) the

TABLE 3

Equilibrium constants for six reactions of technical importance

Reaction	Equilibrium constant					
	At 300 K		At 900 K		At 1500 K	
	Calculated	Observed	Calculated	Observed	Calculated	Observed
$N_2 + 3H_2 \rightarrow 2NH_3$	5.0×10^5	4.4×10^5	1.4×10^{-6}	1.3×10^{-6}	1.8×10^{-9}	2.1×10^{-9}
$CH_4 + Cl_2 \rightarrow CH_3Cl + HCl$	4.5×10^{18}	4.8×10^{18}	4.8×10^6	4.7×10^6	2.2×10^4	2.0×10^4
$2CH_4 \rightarrow HC{\equiv}CH + 3H_2$	2.6×10^{-50}	2.7×10^{-50}	1.8×10^{-8}	2.1×10^{-8}	11	12
$CH_4 + 2H_2O \rightarrow CO_2 + 4H_2$	2.0×10^{-20}	1.9×10^{-20}	2.7	3.1	8.2×10^4	9.8×10^4
$C_2H_4 + \frac{1}{2}O_2 \rightarrow CH_2{-}CH_2{-}O$	2.3×10^{14}	1.5×10^{14}	91	74	0.19	0.27
$CH_4 + NH_3 + 1\frac{1}{2}O_2 \rightarrow HCN + 3H_2O$	5.4×10^{86}	1.0×10^{86}	3.2×10^{30}	2.8×10^{31}	8.0×10^{20}	2.4×10^{20}

TABLE 4

Decarboxylation of vinylacetic acid

	$\Delta H^{\ddagger a}$	$\Delta S^{\ddagger b}$	$K_H/K_D{}^c$	$K_{C\text{-}12}/K_{C\text{-}13}{}^d$
Calculated	46.3	−13.9	2.32	1.03
Observed	39.3	−10.2 ± 2.5	2.87 ± 0.23^e	1.035 ± 0.01^e

^{a} Enthalpy of activation (kcal/mol).
^{b} Entropy of activation (e. u.).
^{c} Kinetic isotope effect (ratio of rate constants at 550 K) for replacement of the carboxy hydrogen atom by deuterium.
^{d} Kinetic isotope effect (550 K) for replacement of the carboxy carbon atom (^{12}C) by ^{13}C.
^{e} Data for PhCH:CH·CH$_2$·CMe$_2$·COOH, the rate of decomposition of which at 550 K is the same as that for CH$_2$:CH·CH$_2$·COOH.

most stable form to be the doubly bridged structure (2) with the triply bridged (1) higher in energy by only 1 kcal/mol. This assignment was supported by a comparison of the observed vibration frequencies with those calculated; most of the observed spectroscopic bands corresponded to (2) with a few additional weaker ones corresponding to (1).

Given the vibration frequencies and geometry of a rigid molecule, the various thermodynamic quantities (entropy, specific heat *etc.*) can be calculated by standard procedures. The entropies found in this way by MINDO/3 calculation agree with experiment to 0.5 e.u. and even specific heats are well reproduced (Dewar & Ford 1977). If then we know the enthalpy of reaction for a given reaction at one temperature, we can use the specific heats calculated by MINDO/3 for the reactants and products to estimate the heat of reaction at any other temperature, and we can then use the MINDO/3 entropies to estimate the entropy of reaction and hence the free energy of reaction and hence in turn the equilibrium constant. Since heats of reaction at 298 K are readily available, this procedure may be of practical value. Table 3 illustrates its application to six reactions of technical importance (Dewar & Ford 1977); the calculated and observed equilibrium constants agree remarkably well over the whole range of temperature from room temperature to 1500 K except for the last reaction. Here larger errors would be expected because MINDO/3 does poorly for cyanides, in particular HCN.

The success of MINDO/3 in this connection suggested that it should prove equally effective in the calculation of entropies of activation and kinetic isotope effects. This could be extremely useful, because comparison of calculated and observed entropies of activation would provide a very good test of predicted transition-state structures and because the current theoretical interpretations of isotope effects leave much to be desired. Calculations for several reactions have led to satisfactory results. A typical example is shown in Table 4, for the concerted intramolecular decarboxylation of vinylacetic acid (Dewar & Rzepa 1977).

One of the more encouraging features of MINDO/3 has been its success in making predictions, some of which seemed surprising at the time but which have been later confirmed experimentally. For example, MINDO/3 predicted (Dewar & Li 1974) *m*-benzyne (3) to be a stable singlet species with a heat of

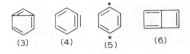

formation similar to that of the well known *o*-isomer (4). Most people at the time would have thought that *m*-benzyne would be a biradical with a triplet ground state, but recent experimental studies (Washburn 1975) seem to have confirmed our prediction. Likewise, *p*-benzyne was predicted to exist not only in a biradical form (5) but also as the isomeric bicyclo[2,2,0]hexatriene (6) which is higher in energy than the biradical (5) but separated from it by an energy barrier so that it should be capable of independent existence. This prediction has also been confirmed (Breslow *et al.* 1975).

A third, and even more startling, prediction (Case *et al.* 1974) concerned the electrocyclic conversion of the cyclobutadiene dimers (7) and (8) into cyclooc-

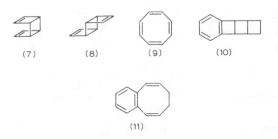

tatetraene (9). According to MINDO/3, the reaction should lead to triplet excited (9), a prediction supported by the formation of the excited product in the analogous conversion of the tetracycle (10) into the bicyclic (11).

A fourth prediction concerned the reaction of singlet oxygen ($^1\Delta_g O_2$) with certain unsaturated hydrocarbons to give epoxides. A general study (Dewar & Thiel 1975) of the reactions of olefins with singlet oxygen had led us to conclude on that, in the absence of $-E$ substituents (e. g. NMe_2, OR), the first product formed is a peroxiran (12) which subsequently rearranges to the observed product. MINDO/3 calculations indicated that such species should react easily with further $^1\Delta_g O_2$ to give the oxiran and ozone. P. D. Bartlett *et al.* (personal communication, 1975) subsequently found that ozone is formed in the reactions leading to oxirans.

Finally, a MINDO/3 study (Bingham *et al.* 1975) of the reaction of the chloride ion with methyl chloride led to the conclusion that the reactants should combine exothermically to form the stable adduct (13) (see p. 116) with pentacovalent carbon, i. e. the structure normally attributed to the $S_N 2$ transition state! After we had arrived at this unexpected result we heard of unpublished work by Dougherty who had found by negative ion mass spectrometry the formation of a stable species $CH_3Cl_2^-$ by combination of Cl^- with CH_3Cl in the gas phase. Table 5 compares our calculated heats of reaction (Dewar & Carrion

TABLE 5

Heats of reaction between chloride ion and chloromethanes

Chloromethane	Product	Heat of reaction (kcal/mol)	
	$Cl^- + RCl$	Calculated (MINDO/3)	Observed[a]
CH_3Cl	$(Cl \ldots CH_3 \ldots Cl)^-$	−12.6	−8.6
CH_2Cl_2	$(Cl \ldots CH_2Cl \cdot Cl)^-$ $^-Cl \ldots H_2CCl_2$	−17.2 −18.3	−15.5
$CHCl_3$	$(Cl \ldots CHCl_2 \ldots Cl)^-$ $^-Cl \ldots HCCl_3$	−13.8 −29.3	−19.1
CCl_4	$(Cl \ldots CCl_3 \ldots Cl)^-$	−14.1	−14.2

[a] Dougherty *et al.* (1974).

1977) for several such processes with those observed by Dougherty *et al.* (1974). In the case of chloroform, and perhaps also dichloromethane, the adduct is of hydrogen-bonded type, but in the other instances it has the trigonal bipyramidal structure. We have also found that other ionic $S_N 2$ reactions are predicted to take place with zero activation in the gas phase, a conclusion apparently supported in several cases by recent ion cyclotron resonance studies.

There is, therefore, good reason to believe in the general correctness of the reaction mechanisms derived from MINDO/3 and MNDO. The picture they give is undoubtedly blurry, owing to their limited accuracy, but probably not distorted. Thus, if there are two possible mechanisms for a reaction and if the calculated activation energies differ by only a few kcal/mol, we shall not be able to say with certainty which of the two will be the easier mechanism, but we are likely to get a reasonably good picture of each of the two transition states.

In the rest of this review I shall describe a few recent applications of MINDO/3 and MNDO which have led to unexpected or interesting conclusions.

```
    0      (25)      25       32      -62
```
relative energy (kcal/mol)   SCHEME 1

The first is concerned with the Diels–Alder reaction. Is this a synchronous process in which both the new bonds are formed to comparable extents in the transition state, or is it a two-stage process, as suggested by Woodward & Katz (1959)? Most chemists currently favour the former alternative, particularly since the work of Woodward & Hoffman (1969) on pericyclic reactions. According to their views, it is difficult to see how such an 'allowed' process could be other than synchronous. MINDO/3, however, does not agree with this. The mechanism it predicts, in the case of the simplest possible Diels–Alder reaction (i.e. that between ethylene and butadiene), is that shown in Scheme 1 with the relative energies (kcal/mol) of the various species; the value in parentheses refers to the transition states (Dewar & Rzepa 1977). The reaction involves the reversible formation of a marginally stable biradical-like intermediate; the main transition state corresponds to conversion of the intermediate into the product. The reaction is a two-step process with one new bond being formed in each step. The calculated energy and entropy of activation agree well with experiment: $\Delta E^{\ddagger}$ calculated 32 kcal/mol, observed 28 kcal/mol; $\Delta S^{\ddagger}$ calculated 9.0 e.u., observed 8.1 e.u.

A recent *ab initio* study of this reaction (Townshend *et al*. 1976) led to the conclusion that it is synchronous, proceeding through a symmetrical transition state. This work is, however, open to all the criticisms implied above, having been done by a method of unknown accuracy without proper optimization of geometries and without proper location or identification of stationary points on the potential surface. This extremely expensive investigation, therefore, led to results of no chemical significance. Arguments by McIver (1972), based on symmetry considerations, make it extremely unlikely that the symmetrical structure can be the transition state in situations such as this, and MINDO/3 predicts it to be much higher in energy than the true unsymmetrical transition state.

Our calculations raise a further interesting point. If the biradicaloid intermediate represents a significant minimum in the potential surface, rotation about the ethylene double bond may occur before it dissociates back into the reactants. In this case, butadiene should catalyse *cis–trans* isomerization of ethylene, a possibility we are currently studying experimentally.

(13)   (14)   (15)

Studies by MINDO/3 and MNDO of several other Diels–Alder reactions indicate that they too proceed by a two-step or two-stage mechanism through an unsymmetrical transition state. This is true even for the reaction of furan (14) with maleic anhydride (15) (Dewar & Ford 1977), a reaction which has been claimed to be synchronous on the basis of deuterium isotope effects. The argument here is that, if the transition state were unsymmetrical with one of the new bonds almost completely formed and the other hardly at all, then deuterium substitution at one, or both, of the relevant carbon atoms (asterisked in 14 and 15) should then be additive, the rate constant (k_1) for the monodeuteriated compound being given in terms of those for the undeuteriated (k_0) and dideuteriated (k_2) species by equation (1).

$$k_1 = \tfrac{1}{2}(k_0 + k_2) \qquad (1)$$

If, however, the transition state were symmetrical, each deuterium atom in either deuteriated species would have an equal effect on the rate, so the isotope effect for the dideuteriated species would be the square of that for the monodeuteriated one; hence the rate constant k_1 would be given by equation (2).

$$k_1 = (k_0 k_2)^{1/2} \qquad (2)$$

The two cases are distinguishable if the rates are measured with sufficient accuracy, and several such studies of symmetrical Diels–Alder reactions have led to the conclusion that the rates follow equation (2) rather than equation (1).

This argument depends, however, on the assumption that the deuterium atoms attached to the weaker forming bond in the transition state will have no effect on the rate. However, our calculations indicate that, although these bonds are still long in the transition states for (ethylene + butadiene) (2.80 Å) and (14 + 15) (2.66 Å), the terminal carbon atoms have already adopted an essentially tetrahedral geometry, so the effect of a deuterium at that position could be just as great as in the other stronger bond. If so, the isotope effects would follow equation (2) even though the transition states are highly unsymmetrical.

TABLE 6

Isotope effects in Diels–Alder reactions

Reaction	Relative rate constants[a]			
	K_2/K_0	K_4/K_0	$\frac{1}{2}(K_0 + K_4)/K_0$	$(K_0 K_4)^{1/2}/K_0$
butadiene + $\mathrm{C}a_2=\mathrm{C}b_2$	0.907	0.820	0.910	0.9055
anthracene + $\mathrm{C}a_2=\mathrm{C}b_2$	0.924	0.852	0.926	0.923

[a] Rate constant for a = b = H, K_0; for a = D, b = H, K_2; for a = b = D, K_4.

To check this, we calculated (Dewar & Rzepa 1977) the isotope effects for reactions of butadiene with ethylene, 1,1-dideuterioethylene ($CH_2=CD_2$) and tetradeuterioethylene ($CD_2=CD_2$). The results, shown in Table 6, conform better to equation (2) than to equation (1), even though the transition state is as unsymmetrical as it could be, one of the new bonds being completely formed while the other is almost double its equilibrium length. Although no experimental data are available for this reaction, data are available (Taagepera & Thornton 1972) for the analogous reactions of ethylene, $CH_2=CD_2$, and $CD_2=CD_2$ with anthracene. As Table 6 shows, they run closely parallel to our calculated values for the butadiene reactions. It had been concluded that the reaction of ethylene with anthracene must be synchronous since the isotope effects follow equation (2) better than equation (1). Our calculations refute this argument, and incidentally indicate that a combination of experimental studies of kinetic isotope effects with MINDO/3 or MNDO calculations may provide a powerful diagnostic tool in studies of reaction mechanisms.

Why are these Diels–Alder reactions non-synchronous? Because they are two-bond processes, i.e. reactions in which two bonds are formed and two broken. Activation is needed in one-bond reactions, where one bond is formed and one broken, because the old bond has to be weakened before the new bond can begin to form. On this basis one would expect the activation energy for a synchronous two-bond reaction to be about double that for an analogous one-bond one. Two-bond reactions, therefore, tend to be non-synchronous, taking place in steps or stages, each of which involves breaking and forming of one bond only.

MINDO/3 and MNDO studies have shown that most two-bond reactions take place in this way. See for example reactions 3 to 7. Reaction (7) is interesting. A recent *ab initio* study (Poppinger 1975) suggested that it is synchro-

[Reaction schemes (3)–(7) shown as structural diagrams]

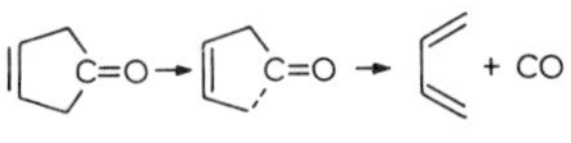

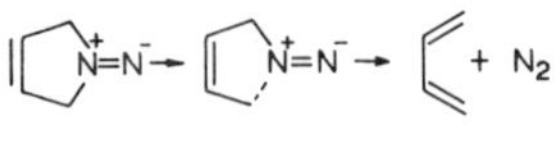

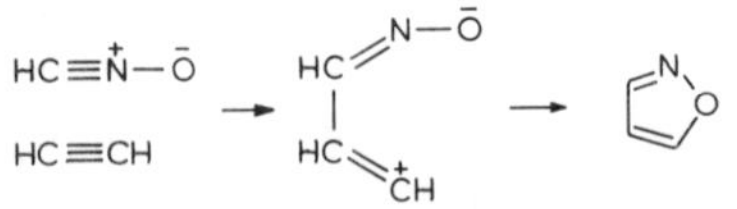

(3) bicyclobutane → biradical → butadiene

(4) norbornadiene → biradical → benzene

(5) cyclopentadienone dimer-type → + CO

(6) diazo bicyclic → → + N_2

(7) $HC≡\overset{+}{N}-\bar{O}$ + $HC≡CH$ → intermediate → isoxazole

nous but MNDO predicts two possible reaction paths, one indicated in equation (7) and the other an alternative two-step mechanism in which the C–O bond is formed first. The transition state for the latter is close in energy to a stationary point on the potential surface that corresponds to the structure expected for a synchronous transition state but which in fact is not, having two negative force constants.

The conversion of bicyclobutane into butadiene, reaction (3), involves a biradicaloid species as a stable intermediate. Rotation about the C–CH$_2$ bond in this needs little activation, so the activation energy for formation of the 'forbidden' product is greater by only 2 kcal/mol than that for the 'allowed' one. Closs & Pfeffer (1968) found that the reaction is stereoselective, not stereospecific, the ratio of 'allowed' to 'forbidden' products implying a difference in activation energy of about 2 kcal/mol. This illustrates another of our conclusions, namely that, although an 'allowed' reaction occurs more easily than an analogous 'forbidden' one if the two processes are otherwise similar, the difference in rate between them need not be large. There is no question of 'forbidden' reactions being impossible, or nearly impossible, as some authors have implied.

Our last example is the Cope reaction (Scheme 2). This has been extensively quoted and discussed as a classic example of a synchronous pericyclic reaction,

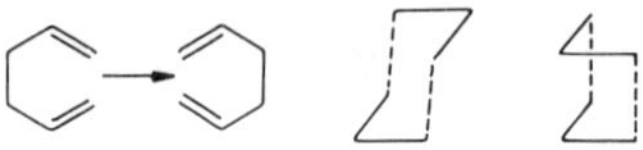

[Scheme 2: 1,5-hexadiene → (16) → (17) chair/boat transition states]

(16)   (17)

Scheme 2

and Doering & Roth (1962) showed that the reaction must, on this basis, proceed through a transition state of chair-type geometry (16) rather than of boat-type (17). On the other hand, McIver (1972) has put forward powerful arguments, based on considerations of symmetry, indicating that the symmetrical intermediate cannot be the transition state, being either a stable intermediate or an unconditionally unstable species.

MINDO/3 calculations confirmed both these conclusions (Wade 1974). There seemed to be two distinct reaction paths, one through a stable symmetrical intermediate (16), the other through a similar stable intermediate (17), and the calculated free energies of activation for the two paths agreed well with experiment (Table 7). For the preferred 'chair' path, the calculated enthalpy and

TABLE 7

Activation parameters for the Cope rearrangement of hexa-1,5-diene

Geometry	$\Delta H^{\ddagger}$ (kcal/mol)		$\Delta S^{\ddagger}$ (e. u.)		$\Delta G^{\ddagger}$ (150 °C)	
	Calculated	Observed	Calculated	Observed	Calculated	Observed
Chair	35.1	33.5	−13.8	−17.0	40.9	40.7
Boat	41.4	44.7	−14.7	−3.0	47.6	46.0

entropy of activation also agree with experiment. The discrepancy in the case of the 'boat' path is almost certainly due to experimental error since the rate measurements were admitted to be less accurate and there is no reason why the entropies of activation for the two paths should differ significantly.

If the rearrangement is a true pericyclic one, involving an aromatic transition state (Dewar 1971), one would expect a 2-phenyl substituent (18) to have little effect on the rate whereas a 3-phenyl substituent (19) should accelerate it significantly, because in (18) the phenyl group is attached throughout to an sp^2-hybridized atom in an even alternant hydrocarbon whereas in (19) it moves into conjugation from a saturated position. We were, therefore, surprised to find (see Wade 1974) that (18) showed a much greater rate increase, relative to hexa-1,5-diene (69-fold at 189.8 °C), than (19) (17-fold). This implied that, as Doering et al. (1971) suggested, the reaction is not a pericyclic one at all but rather takes place via the biradical (20). A further complication was provided by the report by Goldstein & Benzon (1972) who studied the interconversion of the labelled bicyclo[2,2,0]hexanes (21) and (22). This reaction undoubtedly involves fission of the transannular bond to form the labelled biradical (20) formed, at least initially, in the boat geometry (23). If the 'boat' Cope rearrangement also involves this intermediate, one would expect the interconversion of (21) and (22) to be accompanied by the corresponding 'boat' Cope product, (24) or (25). A hexa-1,5-diene was indeed formed, but this proved to be neither (24) nor (25) but (26), i.e. the isomer expected if the boat biradical (23) had to isomerize to the chair biradical (27) before coming apart.

These difficulties have now been elucidated by a more detailed study (Dewar et al. 1977) of the reaction with MINDO/3. Apparently, the potential surface contains a deep crater corresponding to the various conformational isomers of (21), which can thus equilibrate readily. Access to the crater is over one or other of six cols, two leading to hexadiene isomers interconvertible by the chair mechanism, two to hexadiene isomers interconvertible by the boat mechanism, and two to a pair of interconvertible bicyclohexanes (cf. 21 and 22). The course of the reaction is determined solely by the heights of the various cols (transition states) and by the relative stabilities of the products. Here again we have an 'allowed', potentially pericyclic, reaction which is not only not synchronous but not even concerted. The reason again is that the overall reaction is a three-bond process, involving fission of one σ- and two π-bonds and formation of one σ- and two π-bonds. The reaction consequently prefers a two-step path through a stable intermediate.

Even though these examples represent only a tiny fraction of the reactions we have studied, they are sufficient to indicate the potential of our new theoretical techniques. We suspect that use of these in conjunction with experiment will lead to a major advance in our understanding of chemical reactions in organic chemistry and preliminary studies suggest that the impact on organic photochemistry may be still greater. It should also be remembered that MINDO/3 and MNDO are only the first steps in what amounts to an essentially new approach to chemical problems and they have, moreover, been applied so far only to organic chemistry. The possibilities for further development and for extension into other areas of chemistry seem almost unlimited.

ACKNOWLEDGEMENTS

This work has taken a great deal of effort and an immense amount of computer time. It could not have been done without the generous support I have had from the Air Force Office of Scientific Research and the Robert A. Welch Foundation or without the exceptional computing facilities I have had at The University of Texas at Austin. Nor would it have been possible without the efforts of the many able and energetic collaborators I have been fortunate enough to have with me in this endeavour.

References

BINGHAM, R. C., DEWAR, M. J. S. & LO, D. H. (1975) Ground states of molecules. XXV. MINDO/3. An improved version of the MINDO semiempirical SCF-MO method. *J. Am. Chem. Soc.* **97**, 1285–1293

BRESLOW, R., NAPIERSKI, J. & CLARKE, T. C. (1975) The generation and trapping of butalene. *J. Am. Chem. Soc.* **97**, 6275–6276

CASE, R. S., DEWAR, M. J. S., KIRSCHNER, S., PETTIT, R. & SLEGEIR, W. (1974) Possible intervention of triplet states in thermal reactions of hydrocarbons. A study of the rearrangements of cyclobutadiene dimers and analogous compounds. *J. Am. Chem. Soc.* **96**, 7581–7582

CLOSS, G. L. & PFEFFER, P. E. (1968) The steric course of the thermal rearrangements of methylbicyclobutanes. *J. Am. Chem. Soc.* **90**, 2452

DEWAR, M. J. S. (1969) *The Molecular Orbital Theory of Organic Chemistry*, McGraw-Hill, New York

DEWAR, M. J. S. (1971) *Angew. Chem., Int. Edn. Engl.* **10**, 761

DEWAR, M. J. S. (1975) *Chem. Br.* 97

DEWAR, M. J. S. & CARRION, F. (1977), in press

DEWAR, M. J. S. & DOUGHERTY, R. C. (1975) *The PMO Theory of Organic Chemistry*, Plenum Press, New York

DEWAR, M. J. S. & FORD, G. P. (1977) Ground states of molecules. 37. MINDO/3 calculations of molecular vibration frequencies. *J. Am. Chem. Soc.* **99**, 1685–1691

DEWAR, M. J. S. & LI, W.-K. (1974) MINDO/3 study of bisdehydrobenzenes. *J. Am. Chem. Soc.* **96**, 5569–5571

DEWAR, M. J. S. & RZEPA, H. S. (1977), in press

DEWAR, M. J. S. & THIEL, W. (1975) Ground states of molecules. 30. MINDO/3 study of reactions of singlet ($^1\Delta_g$) oxygen with carbon–carbon double bonds. *J. Am. Chem. Soc.* **97**, 3978–3986

DEWAR, M. J. S. & THIEL, W. (1977a) Ground states of molecules. 38. The MNDO method. Approximations and parameters. *J. Am. Chem. Soc.* **99**, 4899–4907

DEWAR, M. J. S. & THIEL, W. (1977b) Ground states of molecules. 39. MNDO results for molecules containing hydrogen, carbon, nitrogen, and oxygen. *J. Am. Chem. Soc.* **99**, 4907–4917

DEWAR, M. J. S. & WADE, L. E. (1977). A study of the mechanism of the Cope rearrangement. *J. Am. Chem. Soc.* **99**, 4417–4424

Dewar, M. J. S., Ford, G. P., McKee, M. L., Rzepa, H. S. & Wade, L. E. (1977) The Cope rearrangement, MINDO/3 studies of the rearrangements of 1,5-hexadiene and bicyclo[2.2.0]hexane. *J. Am. Chem. Soc. 99*, 5069–5073
Doering, W. von E. & Roth, W. R. (1962). *Tetrahedron 18*, 67
Doering, W. von E., Toscano, V. G. & Beasley, G. H. (1971) Kinetics of the Cope rearrangement of 1,1-dideuteriohexa-1,5-diene. *Tetrahedron 27*, 5299–5306
Dougherty, R. C., Dalton, J. & Roberts, J. D. (1974). *J. Org. Mass. Spectrom. 8*, 77
Goldstein, M. J. & Benzon, M. S. (1972a) Inversion and stereoselective cleavage of bicyclo[2.2.0]hexane. *J. Am. Chem. Soc. 94*, 5119–5121
Goldstein, M. J. C. & Benzon, M. S. (1972b) Boat and chair transition states of 1,5-hexadiene. *J. Am. Chem. Soc. 94*, 7147–7149
McIver, J. W. Jr. (1972) On the existence of symmetric transition states for cycloaddition reactions. *J. Am. Chem. Soc. 94*, 4782–4783
Pople, J. A. & Beveridge, D. L. (1970) *Approximate Molecular Orbital Theory*, McGraw-Hill, New York
Poppinger, D. (1975) Concerted 1,3-dipolar addition of fulminic acid to acetylene and ethylene. An *ab initio* molecular orbital study. *J. Am. Chem. Soc. 97*, 7486–7488
Taagepera, M. & Thornton, E. R. (1972) Secondary deuterium-isotope effects—transition-state in reverse Diels–Alder reaction of 9,10-dihydro-9,10-ethanoanthracene—potentially general method for experimentally determining transition-state symmetry and distinguishing concerted from stepwise mechanisms. *J. Am. Chem. Soc. 94*, 1168
Townshend, R. E., Ramunni, G., Segal, G., Hehre, W. J. & Salem, L. (1976) Organic transition states. V. The Diels–Alder reaction. *J. Am. Chem. Soc. 98*, 2190–2198
Wade, L. W. (1974) *Ph. D. Dissertation*, The University of Texas at Austin, Austin, Texas
Washburn, W. N. (1975) Generation of bicyclo[3.1.0]hexatriene. A reactive intermediate. *J. Am. Chem. Soc. 97*, 1615–1616
Woodward, R. B. & Hoffman, R. (1969). *Angew. Chem., Int. Edn. Engl. 8*, 840
Woodward, R. B. & Katz, T. (1959). *Tetrahedron 5*, 70

Discussion

McCapra: After your display of omniscience, I tremble for my experiments!

Dewar: I should be the last to claim that one could replace experiments by these or other techniques but, within the limits of accuracy of these methods, we get results that are good enough to form a better basis for experimental studies than we have previously had.

Woodward: As I recall, you did suggest that these methods be substituted for experiments (Brown *et al.* 1970; cf. Goldstein & Benzon 1972).

Dewar: No! I said that, if we can do the calculations for a reaction, they are worth doing before we study the reaction experimentally. They should give a good indication of what is likely to happen. The trouble with experimental approaches is that one can only disprove various possibilities. We can provide a better basis for trying to *prove* one possibility.

Woodward: What do you find for the triplet–singlet splitting in methylene?

Dewar: By MINDO/3 we find 9 kcal/mol and the MNDO/2 value is about

30 kcal/mol—both are about 10 kcal/mol out from the experimental value. We are less embarrassed than Shaeffer is!

McCapra: If one considers the dissociation of dioxetan, one would expect from what you said about the two-bond process that the biradical would be the transition state. Some evidence (Steinmetzer *et al.* 1974) shows that the excitation process is independent of the transition state whereas you seem to have implied that the transition state is also a transition state for the formation of triplet. What effect would substitution of the dioxetan have on stability?

Dewar: According to (unpublished) calculations it should decrease stability. Electron-releasing substituents should greatly affect it.

McCapra: Thermochemical calculations suggest the opposite. The lifetime for the unsubstituted dioxetan is predicted to be about 10 s at 60 °C and for the tetramethyl compound it is 2.3 h (O'Neal & Richardson 1970). In addition, electron-releasing substituents produce singlets overwhelmingly (F. McCapra, K. A. Zaklika & I. Beheshti, unpublished work, 1977).

Dewar: That is the problem with tetramethyl-1,2-dioxetan. Benson's method of calculating the energy of biradicals is unsound, particularly in this area. His conclusion that the tetramethylene biradical needs activation for cyclization to cyclobutane is certainly incorrect. We have not calculated $\Delta E^{\ddagger}$ for the tetramethyl compound. The average error in our calculations of activation energies is about 6 kcal/mol but it seems likely that we should get better estimates for the relative rates for the substituted and unsubstituted compounds. The question of whether a singlet or triplet is formed is complicated by the possibility of intersystem crossing either way; there is always enough energy. The singlet might be formed by intersystem crossing from the triplet.

McCapra: Also one often does not know the energy levels. However, it is significant that the singlet is found in quantity when the lowest energy state of the carbonyl compound formed is π–π*. Your calculations, understandably, refer to n–π* only and, therefore, one should not extrapolate to other cases.

Dewar: We calculated only the n–π* triplet. Possibly the π–π* surface intersects the n–π* one.

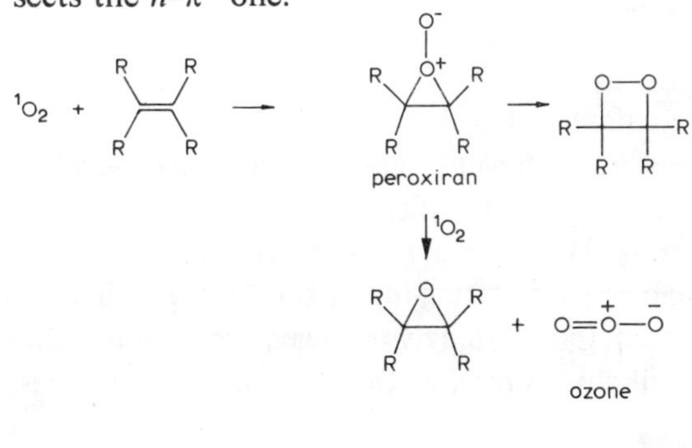

SCHEME 1 (McCapra)

McCapra: Another possibility that has not been considered is rapid intersystem crossing from the singlet $n-\pi^*$ in the solvent cage. Another point refers to the formation of the peroxiran from $^1O_2$ and olefins (e.g. Scheme 1). The products formed depend on the sensitizer (Jefford & Boschung 1976). When we tried the obvious experiment of increasing the illumination (which increases the concentration of $^1O_2$) to see whether more peroxiran was formed, we found that it was not, because the character of the sensitizer overwhelms any minor perturbation like that.

Baldwin: The normal product of the reaction of $^1O_2$ with an olefin containing suitably placed allylic H atoms is the allylic hydroperoxide. Is the peroxiran an intermediate, forming the products by a sigmatropic H shift?

Dewar: Yes. The two-step process is much the more favourable.

Baldwin: We have tried to intercept such peroxirans in olefins which could give the ene-hydroperoxide. Isolable isoelectronic species such as aziridine *N*-oxides and thiiran *S*-oxides (episulphoxides) readily undergo that elimination to the allylic system, but in the dioxygen system we never observed such behaviour. The published experimental evidence for that intermediate is extremely weak.

Dewar: I should add that our MINDO/3 calculations are unusually bad for peroxides since the INDO approximation cannot cope well with lone-pairs, particularly on adjacent atoms, and MINDO is also bad for the oxiran ring system. We found, however, that the rearrangement was facile. The peroxiran is a high-energy species and the barrier to the rearrangement is probably lower than for breakdown to starting compounds.

Cornforth: Some enzymes catalyse the addition of normal O_2 to benzene (equation 1) when the electrons are supplied by an iron–sulphur complex cluster. Both oxygen atoms in the diol come from the same molecule of O_2. Are you in a position to calculate possible mechanisms for such a reaction?

$$C_6H_6 + O_2 + 2e + 2H^+ \longrightarrow \text{cyclohexadiene-cis-diol} \tag{1}$$

Dewar: Yes, but we cannot guarantee that they are right.

Barton: How would you propose to intercept these radicals chemically?

Dewar: First, the biradicals are not proper biradicals because a singlet biradical must be an energy maximum. There are always deformations of some kind which will make it more stable by facilitating interaction between the two radical centres. In a biradical proper, the two singly occupied orbitals have the same energy. Some distortion will always remove this degeneracy and so re-

Ph
⟨structure⟩ → ⟨structure⟩
Ph   Ph
(28)   (29)

duce the energy. So the reactivity will be less than one would expect for a 'real' biradical. We tried the Cope rearrangement of 2,5-diphenylhexa-1,5-diene (28) by heating it in hydroquinone, thinking that the biradical (29) would be more stable than usual for a biradical-like intermediate. In one experiment (before we ran out of material) we thought we had obtained 1,4-diphenylcyclohexane (to judge by mass spectroscopy). However, we do not want to place emphasis on this unless and until the experiment can be repeated.

Woodward: Was it the *cis-* or the *trans-* isomer?

Dewar: I don't know; we didn't identify it further.

Barton: That is a good experiment. Moreover, there are better hydrogen atom donors than hydroquinone.

(30)   (31)

Dewar: W. von Doering's group (personal communication) found that the homologue, 2,6-diphenylhepta-1,6-diene (30), cyclized to the bicycloheptane (31) on being heated.

Rees: In which case, should not the biradical (20) cyclize?

Dewar: No, because the bicyclohexane (32) is higher in energy than hexa-1,5-diene (33); the difference in the heat of formation is about 5 kcal/mol (the

⟨structure⟩ ⟨structure⟩ ⟨structure⟩ ⟨structure⟩
(20)   (32)   (33)   (34)

experimental heat of formation is known). Our calculations erroneously make the bicyclohexane more stable.

Rees: You have encouraged us to believe in the existence of *p*-benzyne (34). So surely the biradical (20) could collapse to give the bicyclohexane (32)?

Dewar: Because of the energy difference there should be only 1 part in 10^4 or 10^5 at equilibrium. The phenyl groups in (30) stabilize the diene form further.

Rees: Would't one expect an enormous difference in the rate of the rearrangement of the 2-phenyl- (18) and the 3-phenyl- (19) compounds, not just a four-fold increase (see p. 119), since they form the benzyl radical, and a non-stabilized secondary alkyl radical, respectively?

Dewar: I must emphasize that the intermediate is not a proper biradical.

Rees: Isn't it at the bottom of a deep energy well?

Dewar: Yes, that is why it is so stable. We have calculated a non-planar geometry for the unsubstituted biradical (35); this results from hyperconjuga-

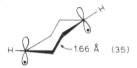

tive interactions between the radical centres (as Hoffman suggested). We should really call these species 'biradicaloid'. They have a closed-shell structure but a small difference in energy between the highest occupied and lowest unoccupied molecular orbitals and are, therefore, very reactive.

Brown: Jones & Bergman (1972) trapped *p*-benzyne with CCl_4. That might also trap cyclohexa-1,4-diyl. If the difference in the heats of formation of bicyclohexane and hexa-1,5-diene is, as you say, less than 10 kcal/mol, then the former might be detected in an isotope dilution experiment.

Dewar: The first step in this reaction is like that in the thermal polymerization of styrene in which a biradical-like species, PhĊH-CH$_2$-CH$_2$-ĊHPh, is formed. Subsequent reaction leads to the initiation process. Here the biradical has been trapped.

Woodward: Does the bond opposite the long 1.66 Å bond (see 35) have the same length?

Dewar: Yes. According to McIver's argument (1972), the symmetrical structure can be either an energy minimum or an energy maximum in two directions, but not a col, which is a maximum only in one direction (along the reaction coordinate). In cases such as this, where two bonds are being formed or broken, the symmetrical structure can be a transition state only if the force constant for stretching one or other bond is smaller than the cross (off-diagonal) constant for interaction between the two stretches. This is very unlikely to be the case because cross-constants are always smaller than diagonal force constants.

Baldwin: We discovered that 2,3-sigmatropic rearrangements of ylides and similar species go by one of two mechanisms depending on the substitution in the precursor: a facile 'concerted' rearrangement or, when the substituents are radical stabilizers, a dissociative recombination. Closs has produced CIDNP evidence for a biradical, not a biradicaloid species. Would two radicals separate in a Cope rearrangement with suitable radical-stabilizing substituents?

Dewar: I'm sure they would. But in a species such as (32) only two stabilizing substituents can be introduced.

Baldwin: In that reaction the bond forms before the biradical. In the reaction I described, it fell apart into two species before any bonds were formed.

Dewar: It gave two separate radicals. There is a qualitative difference between reactions leading to pairs of radicals and those leading to biradicals, because unless the radical centres in a biradical are far apart, stabilizing interactions between them can occur, either across space or through bonds by the hyperconjugative mechanism. We have to be careful in drawing analogies about biradicaloid behaviour from simple radicals.

Breslow: Would you tell us some more about the photochemical work?

Dewar: We have studied the rearrangement of penta-1,4-diene to vinylcyclopropane (Dewar *et al.* 1977). Fig. 1 shows the calculated triplet surface. A general mechanism has been proposed (Zimmerman & Mariano 1969): in the

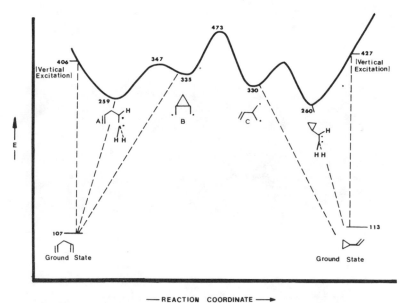

FIG. 7 (Dewar). Energy states (in kJ/mol) for the rearrangement of penta-1,4-diene to vinylcyclopropane via the radical species A, B, C etc. (see Dewar *et al.* 1977).

first step one double bond is excited and rotates to give the biradical-like species A which cyclizes to the cyclic symmetrical biradical B. Breaking a ring C–C bond gives another biradical C which cyclizes to the product. According to Zimmerman & Mariano, once the pentadiene is excited the subsequent steps need little or no activation, whereas our calculations predict an energy barrier of about 140 kJ/mol in the excited state. As a result, ordinary triplet-sensitizers cannot bring about reaction since they do not supply enough energy. Mercury sensitization is effective since it provides 420 kJ/mol. A similar situation may arise in other photochemical reactions. The idea that a photochemical process cannot occur if there is an energy barrier on the excited state surface may not be correct if enough energy is put in. The other reaction we studied was the decomposition of butanal into ethylene and *enol*-acetaldehyde by hydrogen abstraction—the simplest Norrish type II mechanism (see Fig. 2). We find appre-

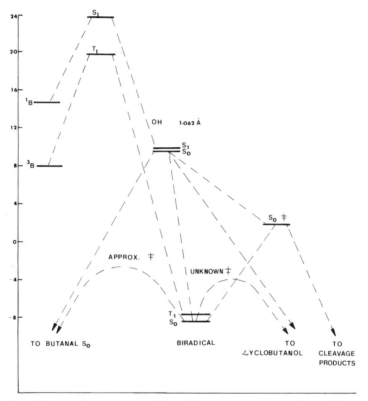

FIG. 8 (Dewar). Energy levels (in kJ/mol) for the decomposition of butanal into ethylene and *enol*-acetaldehyde by hydrogen abstraction.

ciable barriers on both the singlet and triplet surfaces (Dewar & Doubleday 1977). On the triplet surface hydrogen abstraction gives a biradical species and this either returns to butanal or cyclizes to cyclobutanol. Alternatively it can give cleavage products. But the main path for the triplet seems to be cyclization.

For the singlet, the S_0 and S_1 surfaces cross, so one can go directly from the singlet excited state to cyclobutanol or to the cleavage products without going through the biradical. The available experimental evidence suggests that the singlet and triplet reactions differ. The triplet definitely goes through a biradical but the singlet may not. Calculations of this type may prove extremely useful for organic photochemistry because there is little certainty about the mechanisms of such reactions and also because the lowest singlet and triplet states are those mostly commonly involved. What fascinates us in our calculations of ground-state organic reactions is that we keep coming up with mechanisms that are contrary to accepted ideas (e.g. the Cope rearrangement and the Diels–Alder reaction) but are not in disagreement with experiment and on reflection seem indeed more likely than the accepted ones.

Breslow: Do you use multiple electronic configurations (configuration interaction) for the excited state calculations?

Dewar: Triplet surfaces are easy because we can use the unrestricted Hartree–Fock version of MINDO/3 or MNDO without configuration interaction but the singlet surfaces have to be calculated with configuration interaction. It does not need a lot of configuration interaction because we use excited state orbitals. However, inclusion of even a few configurations is enough to increase the computing time by an order of magnitude.

References

Brown, A., Dewar, M. J. S. & Schoeller, W. (1970) MINDO/2 study of the Cope rearrangenent. *J. Am. Chem. Soc. 92,* 5516–5517

Dewar, M. J. S. & Doubleday, C. E. (1977) *J. Am. Chem. Soc.,* in press

Dewar, M. J. S., Kirschner, & Musatto, K. C. (1977) *J. Am. Chem. Soc.,* in press

Goldstein, M. J. & Benzon, M. S. (1972) Boat and chair transition states of 1,5-hexadiene. *J. Am. Chem. Soc. 94,* 7147–7149

Jefford, C. N. & Boschung, A. F. (1976) The dye-sensitized photo-oxidation of biadamantylidene. *Tetrahedron Lett.,* 4771–4774

Jones, R. R. & Bergman, R. G. (1972) *p*-Benzyne. Generation as an intermediate in a thermal isomerization reaction and trapping evidence for the 1,4-benzenediyl structure. *J. Am. Chem. Soc. 94,* 660–661

McIver, J. M. (1972) On the existence of symmetric transition states for cycloaddition reactions. *J. Am. Chem. Soc. 94,* 4782–4783

O'Neal, H. E. & Richardson, W. H. (1970) The thermochemistry of 1,2-dioxetane and its methylated derivatives. An estimate of activation parameters. *J. Am. Chem. Soc. 90,* 6553–6556

Steinmetzer, H.-C., Yekta, A. & Turro, N. J. (1974) Chemiluminescence of tetramethyl-1,2-dioxetane. Measurement of activation parameters and rates of exceedingly slow reactions by a simple and 'nondestructive' method. Demonstration of indistinguishable activation energies for the generation of acetone singlets and triplets. *J. Am. Chem. Soc. 96,* 282–284

Zimmerman, H. E. & Mariano, P. S. (1969) The di-π-methane rearrangement. Interaction of electronically excited vinyl chromophores. Mechanistic and exploratory photochemistry. XLI. *J. Am. Chem. Soc. 91,* 1718–1727

Biomimetic syntheses of phenols from polyketones

G. E. EVANS, M. J. GARSON, D. A. GRIFFIN, F. J. LEEPER and J. STAUNTON
University Chemical Laboratory, Cambridge

Abstract As a result of speculation that many enzymes control polyketone cyclizations *in vivo* by converting a key carbonyl group to a *cis*-enol ether derivative, we describe two novel biomimetic cyclizations. The first involves condensation of two C_6 units derived from triacetic lactone to form an arylpyrone related to aloenin. In the second a naphthapyrone of the rubrofusarin type is formed by condensation of an orsellinic acid derivative with the ether of triacetic lactone.

Biomimetic synthesis (or biogenetic-type synthesis) is a fitting topic for presentation at a symposium held to commemorate the work of Sir Robert Robinson, because it was he who brought home to organic chemists the potential of this line of research. He did so in a spectacular way with his stunningly simple synthesis of tropinone (1) in 1917 while he held the Chair of Organic Chemistry

at Liverpool. The synthesis created a sensation when it was published. One of us (J. S.) remembers how it excited him when decades later he first met it as an undergraduate; and even today it produces a thrill of excitement in each rising generation of organic chemists.

The synthesis can be classed as biomimetic (though the term was unknown to Sir Robert) because its strategy is clearly similar to that of the biogenetic route to the tropane skeleton (in 1917, of course, the biogenesis was still hypothetical). It still ranks as one of the peaks of achievement in this area of endeavour. Since 1917 many chemists have followed Sir Robert's lead particularly in recent years and biomimetic synthesis has be come a major area of research. Workers in the field usually have one of two aims in view. Thus a biomimetic synthesis may be done to test the mechanistic validity of a key step on a hypothetical biogenetic route; in this case it is essential that the chemical model be related as closely as possible to the hypothetical biochemical reaction and the overall synthetic efficiency of the route may be considered of secondary importance. More often, however, the primary objective is the development of an efficient synthesis of a natural product; in this situation the synthetic steps need not be modelled so closely on the biosynthetic steps and the primary considerations in evaluating the results are those usually applied to a synthesis: availability of starting materials and the convenience and overall efficiency of the synthetic scheme. What makes Sir Robert's tropinone synthesis stand out from most other contributions in this field is the fact that it scores full marks on the basis of both biogenetic relevance and synthetic utility.

We shall discuss syntheses that are modelled on the polyketide mode of biosynthesis and so it seems appropriate to begin by recalling Sir Robert's two contributions to the development of biogenetic theory in this area. First, in 1948 he restated the hypothesis, originally put forward by Collie (1907), that poly-

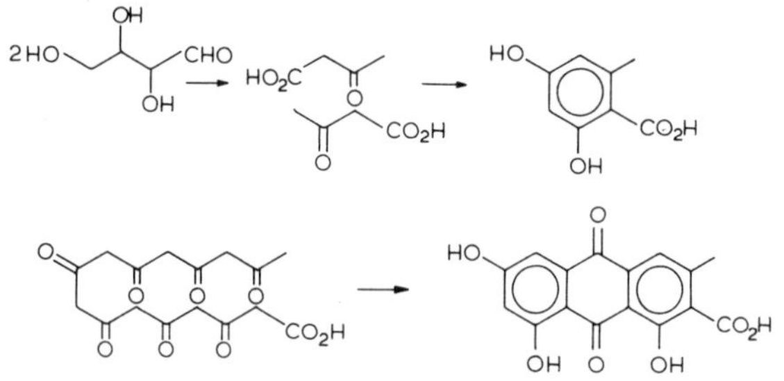

SCHEME 1

phenols arise in nature by cyclization of polyketones. He enlarged on these ideas in 1955 when he proposed that, for example, anthracene derivatives such as endocrocin (2) were also formed by cyclization of suitably folded polyketone chains. In selecting the anthracenoid skeleton for special attention Sir Robert put his finger on the mode of cyclization which would present an exceptional challenge in future years from the biomimetic point of view.

Unfortunately Sir Robert's first publication on this theme appeared in a journal which was not on the mainstream of chemical literature and so it appears to have escaped the notice of most organic chemists. Certainly, the field stayed quiet until the mid-1950s when it was set alight by two papers published by Professor Birch in the *Australian Journal of Chemistry*. In the first, Birch & Donovan (1953) independently proposed the polyketide hypothesis, placing it for

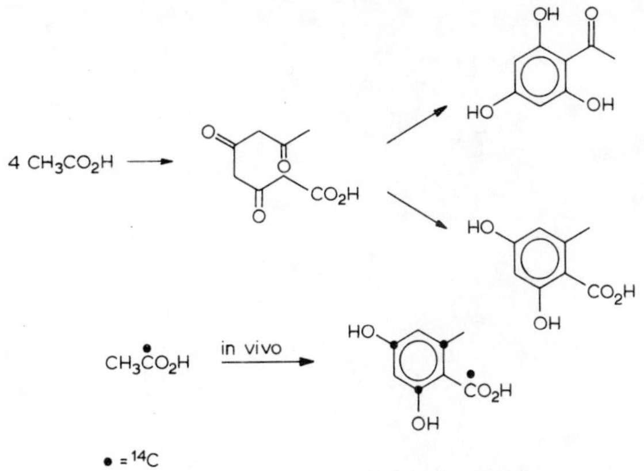

SCHEME 2

the first time on a sound biochemical footing, by drawing attention to the parallel with fatty acid biosynthesis which by that time had been shown to involve linear head-to-tail condensation of acetate units (see Scheme 2). In the second and perhaps even more important paper (Birch et al. 1955), Birch provided the first experimental support for the hypothesis.

With the validity of the hypothesis established there was an explosion of activity on both the synthetic and biosynthetic fronts and both areas of research have remained active to this day. We shall be concerned here solely with developments in the synthetic area and we shall concentrate first on the methods available for the preparation of polyketones and deal later with studies on their cyclization. We shall not mention all the contributions which have been made but give a selection of highlights (for a review, see Harris et al. 1973).

It is fitting that we begin our survey with the polyketone preparation devised by Birch et al. (1963) (Scheme 3). This synthesis consists of ozonolysis of a dihydrobenzene, which was itself generated by Birch reduction of the corresponding aromatic compound. The approach has not been followed up but it

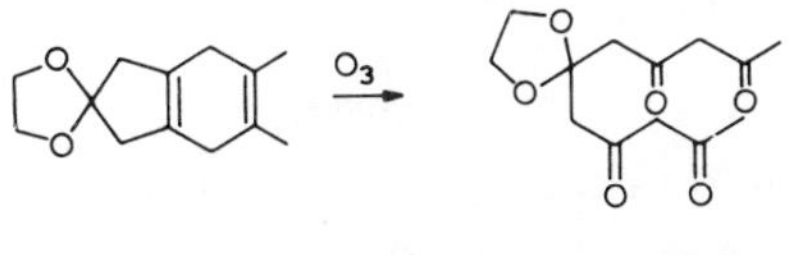

SCHEME 3

may prove important in the future for it offers a method of preparing polyketone chains which are selectively protected at one of the central carbonyl groups.

The second approach (Scheme 4) has its origin in the 19th century when Collie & Myers (1893) investigated the hydrolysis of pyrones to form triketones; it was their observation that these subsequently cyclized to form aromat-

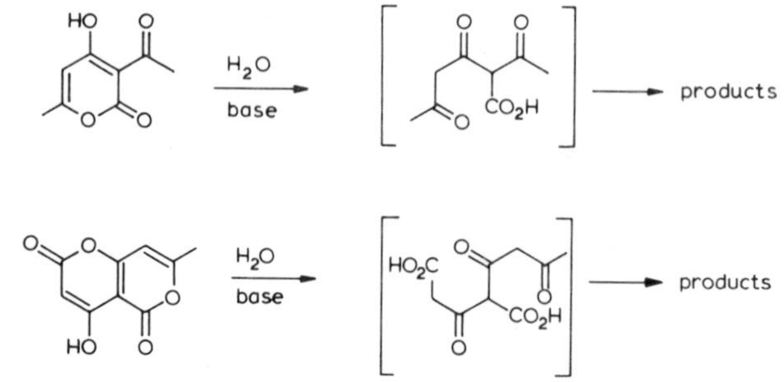

SCHEME 4

ic compounds which led to Collie's biosynthetic speculations. This line of attack has been imaginatively extended by Money et al. (1967) who synthesized complex pyranopyrones such as (3) which is effectively a protected form of the long chain polyketo acid (4). When needed for cyclization studies the polyketo acid can be released as a transient intermediate by hydrolytic cleavage of the oxygen heterocyclic rings.

The most effective approach so far, however, is that developed originally by Hauser & Harris (1958) and since exploited by Harris's group (Scheme 5) (Harris & Murphy 1971; Harris & Murray 1972; Harris et al. 1973). It is based on the discovery that a diketone such as acetylacetone can be converted into a dianion (5) by treatment with strong base. This dianion then reacts with acylating agents at the terminal position to give a linear triketone; the triketone in

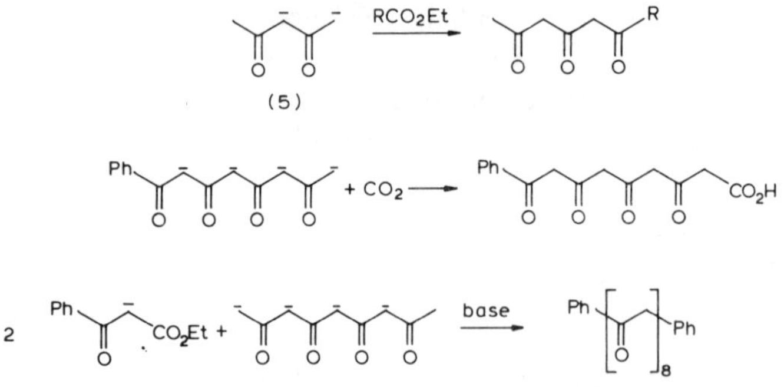

SCHEME 5

turn can be converted via a trianion into a tetraketone and so on. Alternatively, the polyanion can be carboxylated at the terminal position to give a polyketo acid. More recently, linear polyketones with as many as eight carbonyl groups have been prepared by an ingenious adaption of the method in which two successive additions are made at the terminal positions of a polyanion with the monoanion of a β-keto ester as acylating agent. By using variations of these synthetic techniques Harris's group has synthesized an impressively wide range of polyketones, polyketo acids, and polyketo esters.

Given the wide range of compounds at their disposal it is not surprising that most of our information on the behaviour of polyketones on cyclization has come from work done by members of Harris's group. Some representative results showing typical behaviour of triketo acids and esters are given in Scheme 6 (Harris & Carney 1966, 1967; Harris & Murphy 1971). The substrates used in

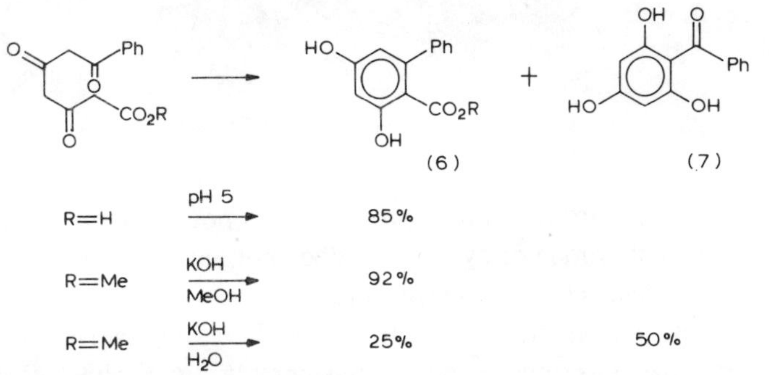

SCHEME 6

these experiments are closely related to the hypothetical polyketones postulated as intermediates in the biosynthesis and so the results are interesting from the biogenetic as well as the synthetic standpoint. As predicted by the biogenetic hypothesis the polyketones can undergo cyclization to form both resorcylic acids (6) and acylphloroglucinols (7). Judged on a synthetic basis these cyclizations are eminently suitable for the preparation of the former but are less successful for the latter. No explanation has been advanced to account for the selectivity shown in these reactions. Why, for example, should the methyl ester prefer to cyclize one way in methanolic base and the other way in aqueous base?

Similar results have been obtained (Money *et al.* 1967) in their studies on the opening and recyclization of pyranopyrones (see Scheme 7). Here the resorcylic acid mode of cyclization was favoured in aqueous base whereas magnesium methoxide in methanol favoured the formation of an acylphloroglucinol. To account for the effect of the magnesium ion Crombie *et al.* (1966; Crombie & James 1966) proposed that the metal chelates with the intermediate polyketo ester causing the chain to fold in a manner favourable to the observed cyclization, as shown in (8). Note that one of the carbonyl groups involved in chelation is part of an ester which is 'extra' to the basic polyketone chain; the

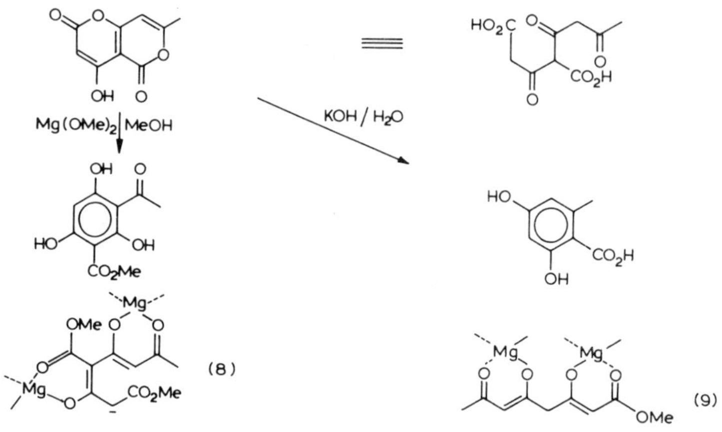

SCHEME 7

equivalent chelate (9) formed from a simple polyketone is not constrained in this way and significantly its mode of cyclization does not appear to be markedly influenced by the presence of magnesium ions.

Thus in the case of triketo acid and ester cyclizations where only two modes of cyclization to form benzenoid rings are open, considerable success has been achieved in controlling the direction taken by the reaction albeit on a largerly hit or miss basis. As the number of carbonyl groups increases so does the number of possible cyclizations and it then becomes necessary to use more direct methods of control.

This is clear from a recent paper by Harris & Wittek (1975) in which they describe studies on hexa- and hepta-ketones. With these longer chains the favoured mode of cyclization is one that involves the terminal carbonyl group leading to the formation of simple benzenoid products rather than a naphthalene or anthracene derivative. Harris & Wittek, therefore, synthesized polyketones in which the two terminal carbonyl groups are protected as acetals. With the favoured cyclization (a in Scheme 8) prevented, the hexaketone der-

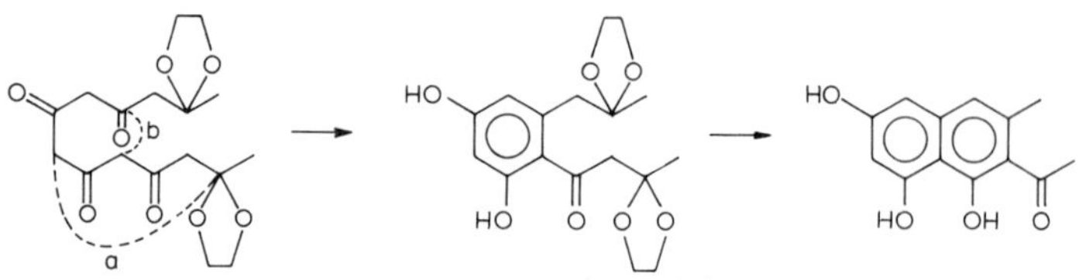

SCHEME 8

ivative (10) can only cyclize in one way (*b*) and so gives (11) (in high yield) which after deprotection cyclizes to the naphthalene (12) in good yield.

This strategem was less successful (Harris & Wittek 1975) with the heptaketone analogue (13). Though one cyclization path (*a*) is blocked, two remain

and of these (*b*) is dominant. Therefore, the intermediate (14), which leads to an anthracene, is formed in no more than trace amounts. The desired product was formed in 10% yield when a more bulky blocking group was used, presumably because of steric hindrance to attack at the neighbouring carbonyl group. Clearly there is scope for improvement here and it will be interesting to see what direction this work takes in the future.

This promising development provides an attractive note on which to end this brief survey of the work done by other groups in the quest for efficient biomimetic syntheses of polyketide metabolites. Our own work in this area began in 1972 as an offshoot of our research into polyketide biosynthetic pathways. We wanted to extend the scope of this research to include certain pathways in which a polyketone cyclizes to form a naphthalene or anthracene derivative. A typical example is the pathway leading to emodin (16) which might, as Sir Robert suggested (1955), involve the anthracene (15) as an intermediate.

To test the validity of hypotheses such as this, we planned to synthesize various polyphenolic naphthalenes and anthracenes carrying isotopic labels at strategic positions. Unfortunately, conventional synthetic routes to such compounds are too cumbersome and inflexible for our purpose. The biomimetic approach on the other hand has a strong appeal for it offers the possibility of inserting an isotopic label at virtually any position in the molecule of interest and in addition it has a straightforward simplicity which holds the promise of convenient and efficient synthetic routes.

With the need for high overall efficiency in mind we decided at the outset to rule out cyclizations of simple polyketones and to aim instead for modified forms of such compounds in which an inbuilt structural feature would direct the molecule towards a particular mode of cyclization.

We argued that the key factor in determining the overall pattern of cyclization is the initial point of folding of the polyketone chain. Once this has been

decided and the first aryl ring has been formed, subsequent cyclizations should proceed spontaneously in the direction shown. Thus starting from the open-chain derivative (17), closure of the first ring (a–a) to form (18) brings into close proximity two polyketone residues which are now ideally placed for cyclization. Three possible modes of cyclization are favoured as a result: c–c, d–d and e–e. The first and third of these are disfavoured by other factors—the former because it requires the formation of a relatively unfavourable anion, the latter because it involves attack at a sterically hindered carbonyl group. Cyclization mode d–d should therefore be dominant and similar factors should direct the formation of the third ring to produce the anthracene (19). Thus, we propose that as each ring is formed it will bring the appropriate parts of the polyketone chains together and so facilitate the formation of the next in line; in effect, the overall transformation bears an interesting resemblance to the closure of a zip-fastener.

This still leaves the problem of controlling the site of the initial ring closure. As the scheme indicates we hope to do this by having a key carbonyl group replaced by a *cis*-enol ether. This device should control the initial cyclization of the chain in the same way that the aromatic rings do in subsequent steps with the consequence that the desired cyclization (*a–a*) will be preferred to the unwanted mode (*b–b*) which Harris found to be preferred in cyclizations of simple polyketones.

Before describing our efforts to realize biomimetic syntheses along these lines we shall discuss the possibility that our chosen strategy may be closely related to the way many enzymes exercise control over the cyclization of polyketone chains. Cyclizations *in vivo* appear to be completely selective. According to above analysis, however, the enzyme that catalyses the formation of an anthracene would not need to possess a formidable battery of strategically placed catalytic groups to achieve this result: having steered the polyketone chain through the decisive steps of the initial cyclization it can then passively allow the molecule to zip up spontaneously, each stage by proceeding by the energetically most favourable mode of cyclization. On this basis one would always expect a linear condensed hydrocarbon skeleton and it may be significant therefore that the naphthalene, anthracene and naphthacene skeletons are commonly represented among polyketide metabolites so far isolated but compounds with a phenanthrene skeleton are extremely rare.

One can continue in this speculative vein to suggest that our proposed strategy for controlling the initial cyclization in biomimetic syntheses—having the key carbonyl group derivatized as a *cis*-enol ether—may also have a parallel in the way many cyclizations are controlled *in vivo*. Again one can find indirect support for the hypothesis this time by surveying the frequency of occurrence of *O*-methylation or reduction at the point of folding of the polyketone chain in various polyketide families. In the family of anthraquinones related to emodin (16) for example, Turner (1971) lists 27 compounds. Of these 11 carry an *O*-methyl group at the site of chain folding against 10 which carry a free hydroxy group at that point. In the remaining six examples, the oxygen function is replaced by hydrogen. This could result from reduction of the carbonyl group to a hydroxy group followed by dehydration to form a *cis*-double bond, which is equivalent to an enol ether in its controlling effect over cyclization. Thus over 60% of the compounds are derivatized at the point of chain folding in a way which could assist the enzyme to direct the cyclization of the first ring. In contrast the hydroxy groups at other positions of the anthracene skeleton of this family of metabolites are only rarely methylated and are never reduced.

A similar pattern can be discerned in many polyketide families in which a polyketone has cyclized to form a naphthalene, anthracene or naphthacene skeleton, but a somewhat different pattern emerges from a survey of metabolites in which a single aromatic ring is generated by cyclization according to the resorcylic acid mode. Here the key hydroxy group is often replaced by hy-

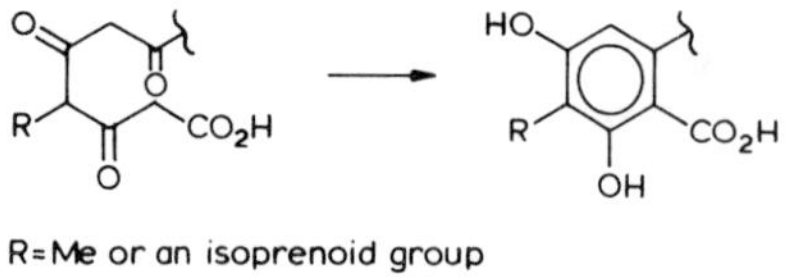

R=Me or an isoprenoid group

SCHEME 9

drogen but it is only rarely methylated. Instead a frequent modification seems to be C-alkylation at the chain folding position (Scheme 9) which raises the possibility that this is another device which can be used by enzymes to control the folding and cyclization of a polyketide chain.

Whether these speculations about cyclization *in vivo* are valid or not, the proposed strategy forms an interesting basis for a biomimetic synthesis and we shall now describe some synthetic work based on it. We are tackling the problems on two fronts. First, we are seeking synthetic routes to *cis*-enol ethers of polyketones so as to study the behaviour of such compounds in cyclization conditions. Secondly, we are investigating the cyclization behaviour of compounds in which a preformed aromatic ring is a potential source of control over the mode of cyclization.

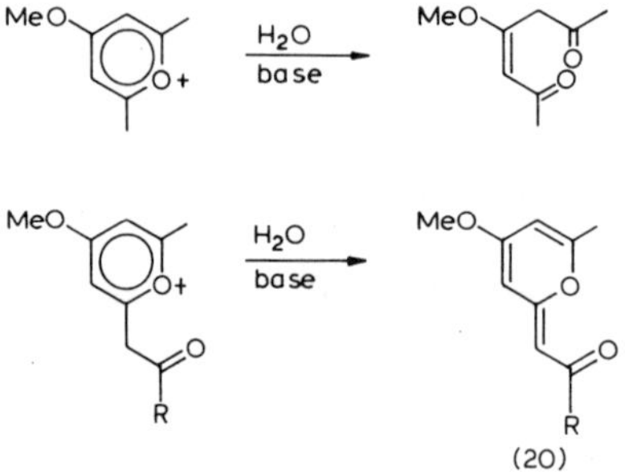

(20)

SCHEME 10

Our first approach to the synthesis of *cis*-enol ethers of polyketones is shown in Scheme 10 (Griffin & Staunton, unpublished results). The pyrylium salts of simple dialkyl-γ-pyrones can be readily opened by base to form the required derivative of a triketone. Unfortunately, however, when a carbonyl substituent is introduced so as to produce the equivalent derivative of a tetraketone, or a triketo ester, the reaction is a deprotonation to give the pyran (20). Such compounds are remarkably stable and they defied all attempts to cleave the heterocyclic ring hydrolytically.

We therefore turned to a different line of attack (M. J. Garson & J. Staunton, unpublished work; Griffin & Staunton 1975). The starting point is the pyrylium salt of an α-pyrone. Such derivatives were not known but we have found that (21) can be readily prepared by treating the corresponding α-pyrone with methyl fluorosulphonate. Treatment of (21) with the anion (22; X=PO(OMe)$_2$)

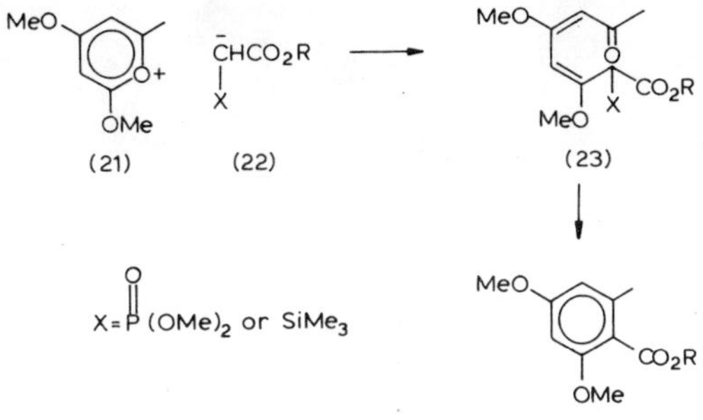

and an extra mole of base (NaH) gives a 65% yield of the resorcylic acid derivative (24) in a 'one-pot' reaction. The transformation is related to a long established reaction of pyrylium salts (Dimroth 1961). Note that the proposed intermediate (23) is a derivative of a triketo acid in which the carbonyl group at the site of chain folding takes the form of a *cis*-enol ether. Attempts to isolate this intermediate were unsuccessful. No other characterizable products have been isolated from the reaction.

In principle this breakthrough could be exploited to produce derivatives of longer polyketone chains by introducing suitable substituents on the alkyl group or by using alternative carbanions in place of (22). We have only explored the latter possibility so far and show in Scheme 11 how we have adapted the process to produce the arylpyrone (25) in three steps from the readily available

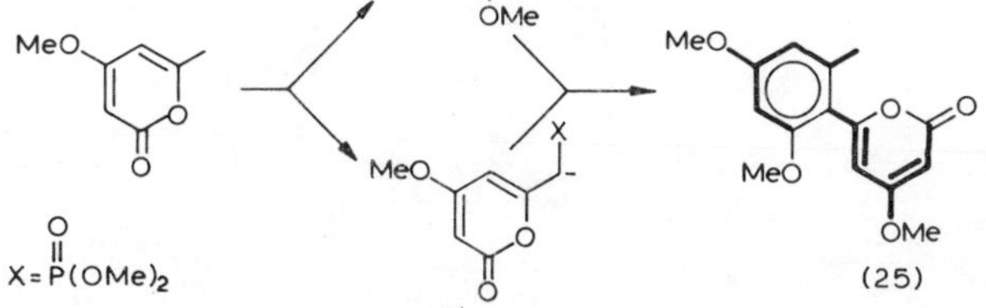

SCHEME 11

enol ether of triacetic lactone. The last step involves the intermediate formation of a derivative of a pentaketo acid (not shown) and so the strategy is basically biomimetic; the skeleton of the product is marked in heavy lines to show how the carbon chain of the intermediate is folded before the formation of the aryl ring. The overall yield is about 20% which means that this biomimetic route is an attractive alternative to conventional syntheses, which start with a preformed aryl ring. An arylpyrone closely related in structure to (25) has recently been isolated from a species of aloe (Suga et al. 1976).

Turning to our alternative strategy in which we plan to use a preformed aryl ring as a means of control in biomimetic cyclizations we have tried one reaction which proved gratifyingly successful. This was modelled on the hypothetical

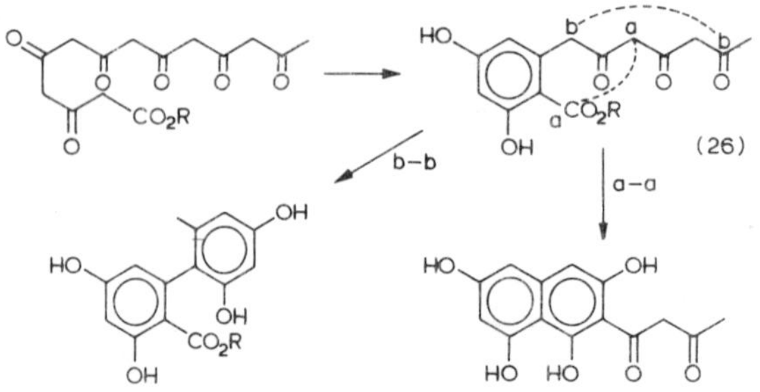

SCHEME 12

biosynthetic transformations shown in Scheme 12. The initial cyclization product (26) is a triketo ester which can undergo two modes of cyclization leading to a naphthalene or a biphenyl (ignoring possible cyclizations onto the ring already present).

Our successful model reaction (G. E. Evans & J. Staunton, unpublished work) involved attack of the anion (27) on the pyrone (28) to form, presumably,

BLOCK 18

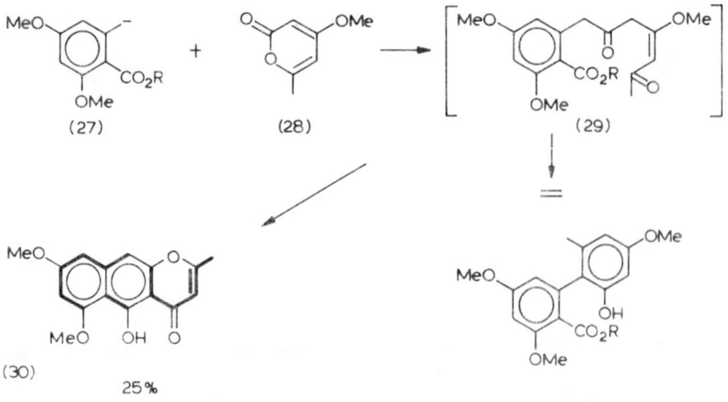

the intermediate (29) which is a derivative of the triketo acid intermediate shown in Scheme 12. This readily cyclizes and the product isolated after treatment with aqueous acid is the naphthapyrone (30). No other compound could be detected in the material recovered from the reaction apart from a considerable quantity of the starting materials so the cyclization appears to be remarkably selective. If our earlier biomimetic synthesis of the orsellinic acid derivative (27) is taken into account the naphthapyrone can be built up from two units of triacetic lactone and one C_2-unit as indicated by heavy lines. Only four steps are involved from the readily available triacetic lactone and the overall yield (15–20%) is excellent, if we bear in mind the complexity of the final product. We wanted to synthesize (30) because the corresponding triphenol is a possible precursor of the fungal metabolite citromycetin. We look forward to putting this biosynthetic hypothesis to the test in the near future.

The challenge, now, is to try to alter the functionality of the triketo acid intermediate in a way which causes it to cyclize to a biphenyl rather than a naphthalene (equivalent to mode *b–b* in Scheme 12). We hope to achieve this result

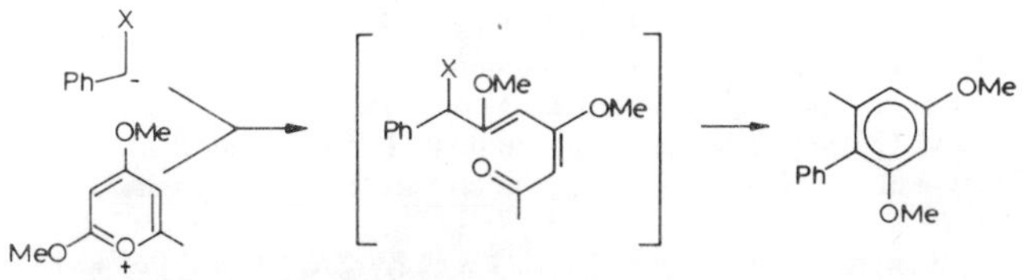

SCHEME 13

along the lines of the successful model reaction shown in Scheme 13 (Griffin & Staunton 1975).

This is just one of many reactions under test at present. Some are bound to fail but even if only a few succeed we shall have made a useful contribution to progress in this area. We like to think that it is a contribution that would have met with Sir Robert's approval.

References

BIRCH, A. J. & DONOVAN, F. W. (1953) Some possible routes to derivatives of orcinol and phloroglucinol. *Austral. J. Chem.* 6, 360–368

BIRCH, A. J., MASSY-WESTROPP, R. A. & MOYE, C. J. (1955) 2-Hydroxy-6-methylbenzoic acid in *Penicillium griseofulvum* Dierckx. *Austral. J. Chem.* 8, 539–544

BIRCH, A. J., FITTON, P., SMITH, D. C. C., STEERE, D. E. & STELFOX, A. R. (1963) Studies in relation to biosynthesis. Part XXXII. Preparation, spectra and hydrolysis of poly-β-carbonyl compounds. *J. Chem. Soc.*, 2209–2216

COLLIE, J. N. (1907) Derivatives of the multiple keten group. *J. Chem. Soc. 91*, 1806–1813
COLLIE, J. N. & MYERS, W. S. (1893) The formation of orcinol and other condensation products from dehydracetic acid. *J. Chem. Soc. 63*, 122–128
CROMBIE, L. & JAMES, A. W. G. (1966) The control of pyrone and aromatic cyclisation in polyketonic-polyenolic systems by magnesium alkoxide concentration. *Chem. Commun.*, 357–359
CROMBIE, L., GAMES, D. E. & KNIGHT, M. H. (1966) Base-catalysed cyclisation of highly enolisable systems: diversion of pathway by magnesium chelation. *Chem. Commun.*, 355–357
DIMROTH, K. (1961) Aromatische Verbindungen aus Pyryliumsalzen, in *Neuere Methoden der Präparativen Organischen Chemie*, vol. 3 (Foerst, W., ed.), pp. 239–260, Verlag Chemie, Heidelberg
GRIFFIN, D. A. & STAUNTON, J. (1975) A novel biogenetic-type synthesis of an orsellinic acid derivative. *J. Chem. Soc. Chem. Commun.*, 675–676
HARRIS, T. M. & CARNEY, R. L. (1966) Biogenetically modeled synthesis of β-resorcylic acids. *J. Am. Chem. Soc. 88*, 2053–2054
HARRIS, T. M. & CARNEY, R. L. (1967) Synthesis of 3,5,7-triketo acids and esters and their cyclizations to resorcinol and phloroglucinol derivatives. Models of biosynthesis of phenolic compounds. *J. Am. Chem. Soc. 89*, 6734–6741
HARRIS, T. M. & MURPHY, G. P. (1971) Synthesis of 1,3,5,7,9-pentacarbonyl compounds. *J. Am. Chem. Soc. 93*, 6708–6709
HARRIS, T. M. & MURRAY, T. P. (1972) Negatively charged electrophiles. Acylation of strong nucleophiles by enolate salts of β-keto esters. *J. Am. Chem. Soc. 94*, 8253–8255
HARRIS, T. M. & WITTEK, P. J. (1975) Biogenetic-type syntheses of polycyclic polyketide metabolites using partially protected hexa- and -heptaketones. *J. Am. Chem. Soc. 97*, 3270–3271
HARRIS, T. M., HARRIS, C. M. & HINDLEY, K. B. (1973) Biogenetic-type syntheses of polyketide metabolites, in *Progress in the Chemistry of Organic Natural Products*, vol. 31 (Herz, W., Grisebach, H. & Kirby, G. W., eds.), pp. 217–282, Springer Verlag, Vienna
HAUSER, C. R. & HARRIS, T. M. (1958) Condensations at the methyl group rather than the methylene group of benzoyl- and acetylacetone through intermediate dipotassio salts. *J. Am. Chem. Soc. 80*, 6360–6363
MONEY, T. F., COMER, F. W., WEBSTER, G. R.B., WRIGHT, I. G. & SCOTT, A. I. (1967) Pyrone studies. I. Biogenetic-type synthesis of phenolic compounds. *Tetrahedron 23*, 3435–3448
ROBINSON, R. (1917) A synthesis of tropinone. *J. Chem. Soc. 111*, 762–768
ROBINSON, R. (1948) The structural relations of some plant products. *J. R. Soc. Arts 96*, 795–808
ROBINSON, R. (1955) *The Structural Relations of Natural Products*, pp. 4–11, Clarendon Press, Oxford
SUGA, T., TORI, K., HIRATA, T. & KOSHITANI, O. (1976) Carbon-13 NMR spectral studies of aloenin and its derivatives. *Tetrahedron Lett.*, 1311–1314
TURNER, W. B. (1971) *Fungal Metabolites*, pp. 156–161, Academic Press, London & New York

Discussion

Birch: The enzymic cyclizations are all-or-nothing processes. Mutation experiments are not very useful in that no released intermediate has been identified. For instance, *Penicillium icelandicum* according to Gatenbeck makes either anthraquinones or nothing from acetate except occasionally when mutants make substituted phthalic acids. These might result from degradation of a compound produced by closure of the first ring of the polyketide chain. Polyketide chains which are not closed are probably reversible to malonyl-CoA.

Unless a key cyclization step intervenes only acetyl-CoA is left and diverted in other directions with no polyketide products.

However, even if cyclization normally operates like a zip-fastener through the chain, there must be a sequence of cyclization steps which must be subject to reactivity considerations. In one speculative mechanism which I suggested for the formation of the larger molecules such as anthraquinone derivatives the outside ring closes first; the other ring closures then follow automatically. A convincing example of the consequences of such initial ring-closure is the lactone, curvularin (31), which undergoes a Claisen condensation in aqueous alkali at room temperature, conditions which are most unusually mild for a C-acylation. Another speculative mechanism for the required ring closures invokes reaction of enol-phosphates with alkali. To obtain a specific product an enzyme would have to select the right carbonyl group to attach the phosphate to.

Staunton: Cyclization of the large ring first would lead to the same conclusions as the zipping up process; on mechanistic terms it is just as attractive.

Kenner: The hypothesis of purely chemical steps after initial enzyme-catalysed reaction gives hope that intermediates may be trapped, unless their steady-state concentration is too low.

Staunton: There is a risk that the trapped intermediate might stay bound to the enzyme and so block the active site.

Barton: To what extent, and where, are these poly(β-diketones) enolized?

Staunton: It varies from compound to compound. Most compounds have at least one enolized carbonyl group; many have several. The n.m.r. spectrum usually indicates the presence of a complex mixture of differently enolized forms.

Barton: The u.v. spectra might be more informative; u.v. absorption is a sensitive indicator of enolization of ring systems. If your target carbonyl group were enolized you could simply methylate it with diazomethane.

Staunton: The risk with that approach is that complex mixtures could be formed. Also there would be a tendency to form *trans-* rather than *cis-*enol ethers.

Barton: That could be overcome by photoisomerization.

Battersby: Dr Staunton, you speculate that the chain is held on the enzyme after methylation of the carbonyl to the *cis*-enol ether. When does that methyl group appear in the natural product?

Staunton: The timing of the methylation is crucial. However, there is no evidence from biosynthetic studies to show whether this step occurs before or after the formation of the first aryl ring. We hope to test our ideas by studying this aspect of the biosynthesis.

Baldwin: Although you say that aromatization of the first ring directs the cyclization to linear annulated systems, the same result would ensue from enolizations on one side only of the folded chain precursor, in other words from the *trans*-enols.

Staunton: I hesitate to endorse that idea because the evidence suggests that these polyketones enolize randomly *in vitro*. The proposal that the enzyme holds the chain in such a configuration for that pattern of enolization seems to imply a battery of directing groups on the enzyme. The concept of one active site directing the formation of one ring with subsequent spontaneous cyclization appeals to us as being much more economical. We intend to investigate the cyclization of such enol-ethers *in vitro*.

Baldwin: Many of the all-*trans*-enols are stabilized by metal ions.

Staunton: The enzyme might contain a row of metal ions.

Dewar: Is the whole molecule formed from the polyketone chain or is an intermediate single-ringed compound first formed which subsequently cyclizes?

Staunton: That is an alternative scheme; it would mean cyclization of a chain such as (32) to give the benzenediol (33). The short chain would then have to be extended by a C_2 unit for the next cyclization to give naphthalene derivative (Scheme 14) and so on. The disadvantage of this scheme is that 2,6-

disubstituted benzoate esters are extremely hindered towards attack by nucleophiles but that cannot be ruled out at present.

Barton: These closures could be directed in another way, namely by carboxylation, either during the biosynthesis or as an extra step. A carboxy group is inserted between the two ketones—a reversible process—and then the carboxylated anion cyclizes.

Staunton: The enzyme would need to have an additional active site to bring about the carboxylation; and unless there were more than one active site the substrate would have to be shifted on the enzyme before each cyclization step.

Barton: The carboxy group could be put in afterwards: the polyketone chain would be carboxylated, the product would cyclize and CO_2 would be liberated with the H_2O.

Birch: There is no evidence for that.

Golding: In the biosynthesis of polyketides, acetyl-CoA combines with malonyl-CoA units. This could provide directly a suitable carboxylated intermediate.

Staunton: In the development of these polyketone chains decarboxylation is an essential part of the C–C bond forming process. Even if it isn't kinetically necessary, decarboxylation has thermodynamic advantages, driving the cyclization in the desired direction. The formation of acetocetate from two units of acetyl-CoA is thermodynamically unfavourable. For that reason, apparently, nature uses malonyl-CoA. That probably rules out the presence of carboxy groups in the lower chain in the later stages.

Dewar: Can the plants use 'unfinished' compounds such as (33) to make anthraquinones?

Staunton: There is no evidence that such compounds can serve as biosynthetic intermediates.

Kenner: Failure in that case might be due to problems of permeability.

Ramage: Also, the polyketide might still be bound to the enzyme at a late stage of its biosynthesis.

Staunton: Simple precursors are not incorporated. The longer ones have not been tried yet.

Birch: Theoretical intermediates fed to the plant are incorporated but only after being degraded into acetyl-CoA.

Ramage: This folding and cyclization of the polyketide parallels the cyclization of squalene epoxide; the enzyme opens the epoxide and then cyclizes it to the steroid ring system. Organic chemists have not been able to control this process entirely in that the thermodynamically more stable tricyclic tertiary carbonium ion intermediate is produced. It may be that the template dictates the mode of cyclization here, too.

Selective homogeneous and heterogeneous catalysis

J. M. BROWN

The Dyson Perrins Laboratory, University of Oxford

Abstract Catalysis of organic reactions by micellar aggregates in aqueous solution lacks synthetic utility in its present state, largely for two reasons: first, the selectivity attainable in a loose unstructured micelle is low, with little possibility of attaining the degree of substrate differentiation possible in enzymes, and, second, the scale of reaction is limited, since surfactant must be in large excess over reactant to avoid swamping of the catalytic effect.

The discussion will cover initial attempts to overcome these drawbacks. Increased selectivity is apparent in the promotion of ester hydrolysis by micelles of asymmetric chain-functionalized surfactants. In the most favourable circumstances 3:1 discrimination in the rate of reaction of ester enantiomers may be observed. Increased scale of reaction is obtainable when polymer-linked cationic surfactants are used as insoluble catalysts, and these hold promise for various anion-activation reactions.

Specific attention is given to mechanistic aspects of homogeneous and heterogeneous catalysis.

Much has been written about the ability of enzymes to catalyse organic reactions with high efficiency and extreme selectivity. Although a detailed discussion of mechanism is inappropriate here, the basic principles are moderately well understood. The tight binding of reactive groups and substrate reaction centre is perhaps the most important factor, leading to minimal atomic motion on the approach to transition-state (Jencks 1975). In addition, the reaction components are transferred from water to a desolvated environment with consequent enhancement in reactivity. This latter point is an important one, since an organic chemist is apt to avoid reactions in water, and yet nature has dispensed entirely with the need for ether, chloroform and dimethyl sulphoxide. Even without catalysis the unique structure-making effect of water may be advantageous (witness the selective terminal addition to squalene in water-rich solvents) and this factor is crucial to the structural organization of enzymes.

In a loose way, one might consider an enzyme–substrate complex as having unidimensional mobility, with complete motional freedom only in the direction of the reaction coordinate. This contrasts with the complete three-dimensional mobility and consequent random collisional orientation of a reacting pair of molecules in an isotropic fluid medium. It further suggests that bimolecular reactions at surfaces and interfaces, where molecules and ions effectively have only two-dimensional mobility, may have different characteristics. In the particular case of aqueous interfaces, this might mean enhanced reaction selectivity when compared to its corollary in dilute solution. Regrettably, this possibility has barely been explored. In a classic example, erucic acid was shown

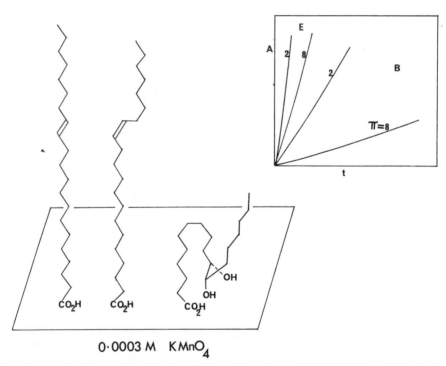

FIG. 1. Selective oxidation of erucic acid (E) and brassidic acid (B) monolayers by 0.3mM-KMnO₄. Inset: rate of reaction (A = surface area/molecule, which increases during the course of reaction; π = surface pressure, in dyne/mm).

to be more rapidly oxidized in monolayers than brassidic acid (Fig. 1) (Rideal 1945; Rideal & Davies 1961). More recently, and particularly owing to the efforts of Whitten and his co-workers (Sprintschink *et al.* 1976; Hopf & Whitten 1976; but cf. Valenty & Gaines 1977), photoreactions of hydrophobic metal-ion complexes in monolayers have been shown to proceed by different paths than similar reactions in solution (see Fig. 2). This is clearly an open field with exciting prospects.

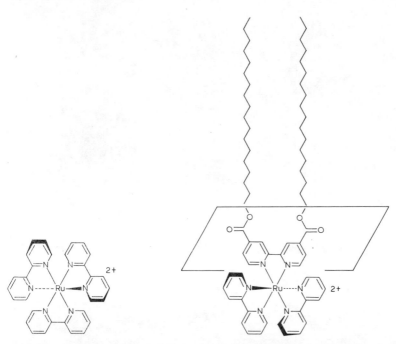

FIG. 2. Photoxidation of water by metal complexes at aqueous interfaces: for $Ru^{II}(bipy)_3$ (left) electron-transfer in aqueous solution is slower than excited-state decay; in its excited state $Ru^{II}(bipy)_3$ (right) can photoreduce H_3O^+ (monolayer luminescence 660 nm, quenched by H_2O, H_2 and O_2 evolved).

Insoluble monolayers provide an extreme case of the general phenomenon that hydrophobic and particularly amphiphilic organic compounds may have substantial surface-excess concentrations. Many reactions thought to occur by association-prefaced catalysis at high dilution may include a contribution from reaction in the surface-excess layer. A striking example is the reaction between decylamine and p-nitrophenyl decanoate (Oakenfull 1973) which is rapid, with a substantial contribution from a third-order term of the form $k_3[C_{10}H_{21} \cdot NH_2]^2[C_9H_{19} \cdot CO \cdot O \cdot C_6H_4 \cdot NH_2]$, this being around 10^7 times faster than the corresponding third-order component of reaction between ethylamine and p-nitrophenyl acetate (see Fig 3). Back-of-envelope calculations indicate an intrinsic *possibility* that the bulk of these hydrophobic reactants is present in the surface-layer. Whether this is an important contributor to catalysis has yet to be determined.

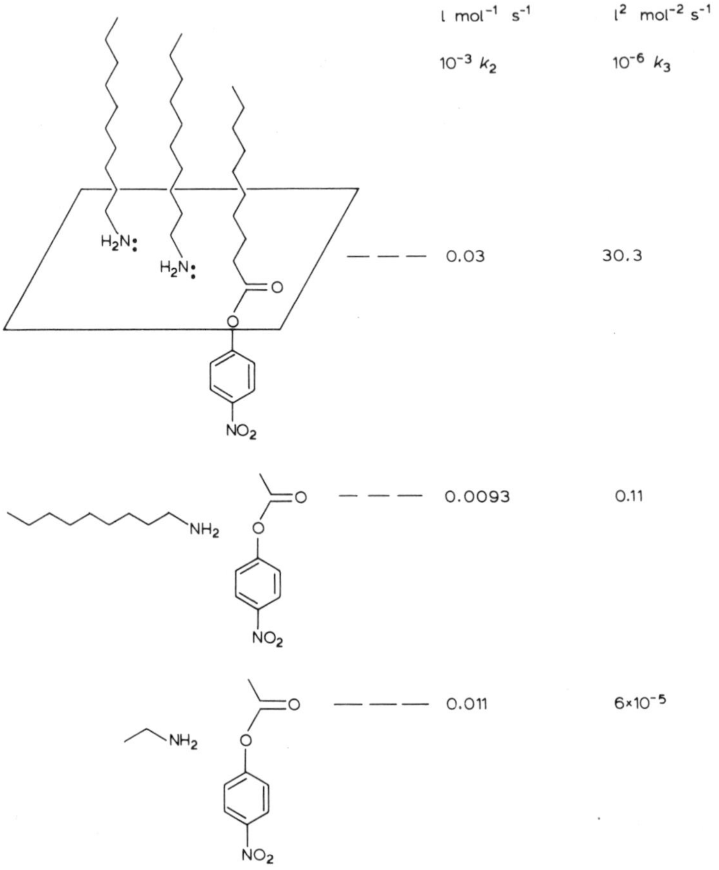

FIG. 3. Association-prefaced catalysis: if one assumes that p-nitrophenyl decanoate has a molecular surface area of 0.2 nm², there are 8.3 × 10⁻¹⁰ mol/cm² of saturated surface. Hydrolyses are done in 5μM-solution.

In more concentrated aqueous solution, amphiphilic molecules aggregate to form micelles. This is particularly true for compounds in which a long alkyl chain is bound to a polar or charged head-group. In dilute solution, the aggregates typically contain about 100 monomers and are almost spherical. The concentration regulating the onset of micelle formation depends on several factors, particularly the length of the alkyl chain and the nature of the head-group, but there is a sharp break in colligative properties associated with that concentration, and the aggregates have a narrow range of dispersisity. Micelles are rapidly formed and broken down, and it is only recently that their dynamic properties have been defined (Aniansson *et al.* 1976) or that useful theories to rationalize their shape and structure have become available (Israelachvili *et al.* 1976). Relaxation kinetics have demonstrated that in simple cases, the micelle/monomer association occurs at a rate close to diffusion control, and complete breakdown of the aggregate may occur on a millisecond time scale. The currently accepted structural theory is based on the fact that an interface to water has its lowest free energy when the aggregate is spherical. With increasing surfactant concentration this may lead in turn to spherical, toroidal and rod-shaped micelles (see Fig. 4).

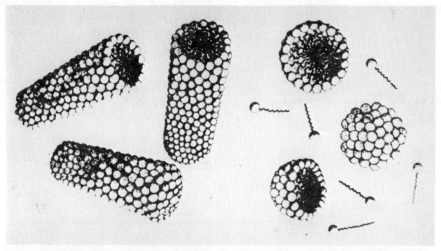

FIG. 4. The structure of micelles: spherical (right) and rod-like (left) micelles are illustrated.

Although the analogy between a loosely organized, rapidly-dissociating micelle and a highly structured enzyme is weak, there are nevertheless some similarities. The most obvious is the structural separation of a hydrophobic core from a hydrophilic exterior in both cases. Of more direct chemical relevance is the ability of micellar aggregates to catalyse hydrolytic and other reactions in aqueous solution (see Fig. 5). Several factors may contribute to this apparent catalysis, which is most pronounced when the surfactant head-group and reaction transition-state are of opposite charge. Briefly, these may be summarized:

(a) the reactants approximate on binding within the micelle, thereby increasing their effective bimolecular concentration;

(b) the transition-state has lower free energy than in aqueous solution because of favourable Coulombic forces;

(c) charged reagents bound to the micelle may be desolvated, and therefore are more reactive;

(d) the micelle may exert a favourable medium effect on the reaction, being of lower polarity than the surrounding water.

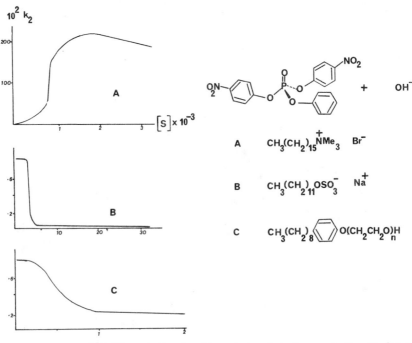

FIG. 5. Micellar catalysis (illustrated by plot of k_2 against surfactant concentration, S) of bis (*p*-nitrophenyl) phenyl phosphate by base with (A) trimethyldecylammonium bromide, (B) sodium dodecyl sulphate, and (C) *p*-octylphenyloxypolyethoxyethanol (from Fendler & Fendler 1975).

The net catalysis which may result from a combination of these factors is modest — less than 10^4 in terms of rate enhancement for the most favourable cases. For our purposes, it is important to note that the Stern layer serves as reaction environment, and hence micellar catalysis may be considered to be a type of surface reaction, although the surface is rough and irregular. By analogy to the earlier discussion, the reactants therefore have two-dimensional freedom and this may allow the attainment of selectivity not accessible in dilute solution.

Our own interest has been in studying the chemistry of, and catalysis by, surfactants possessing a functional group bonded to the long alkyl chain. Much of the synthesis is based on *C*-alkylation and hydrolysis of cyclic lactams, giving rise to chain-carboxylated quaternary ammonium salts which may in turn be converted into, for example, imidazole or hydroxamic acid derivatives. I shall give three examples which demonstrate different aspects of their chemistry.

STRUCTURE OF FUNCTIONAL MICELLES

An important aspect in any study of catalysis must be the structure of the catalyst, and micelles are rather ephemeral, for reasons outlined above. Nuclear magnetic resonance spectroscopy, particularly ^{13}C, is valuable but such measurements must be done in moderately concentrated solution (about 0.4 mol/l). One type of probe is the effect of paramagnetic ions such as Gd^{3+} or moderately stable free-radicals on the spectra. Such techniques were pioneered at Oxford

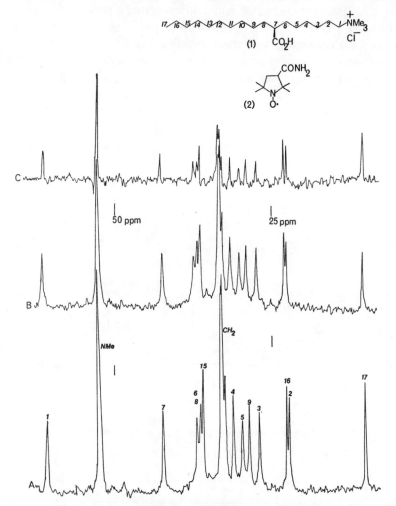

FIG. 6. Effect of a paramagnetic probe on the structure of chain-carboxy micelles: (A) ^{13}C n.m.r. spectrum of 0.5M-7-carboxyheptadecyltrimethylammonium chloride (1) in D_2O; (B) as (A) but with 0.01M-(2) added; (C) difference spectrum.

for the study of enzymes (Campbell *et al.* 1973). Proton decoupling in pulsed ^{13}C n. m. r. spectroscopy results in a Nuclear Overhauser effect which, with appropriate experimental conditions, enhances all carbon absorptions to the same degree and the peak areas accurately reflect the number of carbon atoms associated with the resonance. The presence of a paramagnetic species partially destroys this effect at concentrations well below those resulting in the broadening of the peak. Since the extent of relaxation bears an r^{-3} dependence on the distance between the probe and carbon atom, it should be possible to map the structure of an aggregate and define whether a particular atom is close to the surface or buried deep in the micellar core. The spectra obtained emphasize the random character of these micelles (Fig. 6).

A second type of experiment requires the measurement of ^{13}C spin-lattice relaxation times for carbon nuclei in micelles. This provides information about the mobility of individual carbon atoms since dipole–dipole relaxation, the major mechanism for protonated carbon atoms, increases as molecular motion slows to approach the Larmor frequency. The spectra of 7-hydroxyoctadecyltrimethylammonium iodide serve as an example (Fig. 7). Micellar mobility is

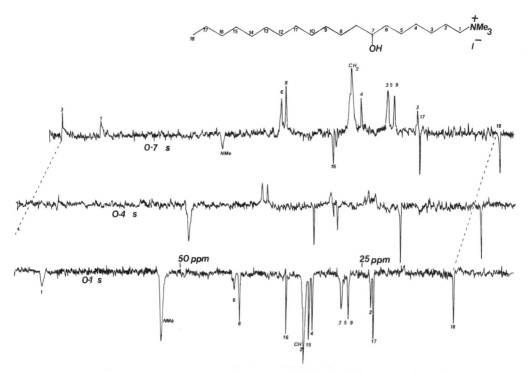

FIG. 7. Spin-lattice relaxation times in micelles (0.4 mol/l, 353 K, 180-τ-90 – T).

moderately constant along the chain, save in the region of the polar hydroxy group. It seems reasonable to postulate that intermolecular hydrogen-bonding, both to other surfactants and to micellated water, is responsible. Additional experiments with paramagnetic counter-ions establish that the hydroxy group, on average, is buried in the micelle remote from the Stern layer (Brown & Schofield 1975).

It is worth emphasizing that these and other experiments demonstrate that the micellar interior is liquid-like. Although there is some reduction in mobility, particularly in the region of polar functional groups, the level of organization is low, and individual surfactant molecules are in rapid dynamic equilibrium.

ASYMMETRIC CATALYSIS (see also Brown & Bunton 1974)

The branced-chain histidine derivative (1) (see Fig. 8), synthesized from caprolactam, contains two asymmetric centres. In its preparation, a single crystalline salt was isolated which had the characteristics of a single diastereomer, most notably a ^{13}C spectrum with the expected number of resolved peaks and

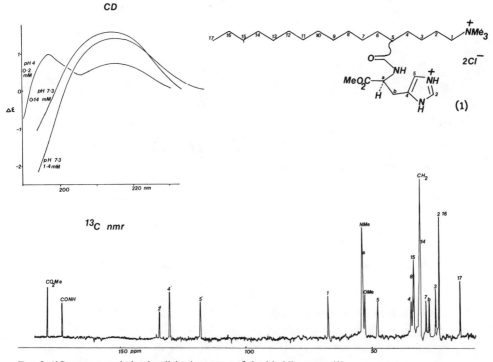

FIG. 8. ^{13}C n. m. r. and circular dichroic spectra of the histidine ester (1).

no evidence of pairwise splittings. Hydrolysis in 6M-HCl gave the 5-carboxylic acid with negative circular dichroism at short wavelength.

This surfactant proved to be effective in the hydrolysis of p-nitrophenyl esters, with variable chiral discrimination for those derived from enantiomeric acids. Hydrolyses were done in aqueous phosphate buffer solution at pH 7.3, in a region where the neutral imidazole was shown to be the reactive species. The rate increases linearly with pH above 8, for here the imidazolide anion contributes to the overall process.

The best asymmetric induction corresponds to an effective enantiomer excess of 51%. Although modest, one needs to understand its mechanistic origins to make progress. This first requires detailed knowledge of the catalytic mechanism, and several precedents demand that the rate-determining stage be nucleophilic rather than general-base (see, in addition, the discussion below). The solvent isotope effect k_{H_2O}/k_{D_2O} is 1.4 in the neutral pH region for the R enantiomer and 1.3 for p-nitrophenyl β-phenylacetate (see Fig. 9).

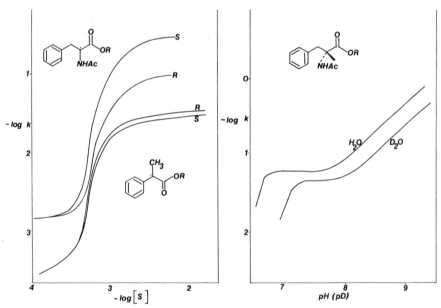

FIG. 9. Relative reactivity of enantiomeric p-nitrophenyl (R) esters with the histidine ester (1).

We synthesized compounds (2) and (3) (see Fig. 10) to test the effects of varying the histidine orientation in the micelle. In neither case was a clear separation of one diastereomer effected; (2) consisted of liquid crystalline microspheres and (3) was a thick gum with no pretensions to crystallinity. The acyl

SELECTIVE HOMOGENEOUS AND HETEROGENEOUS CATALYSIS

region of their ^{13}C n. m. r. spectra (Fig. 10) demonstrates in each case the presence of a pair of diastereomers, in contrast to the sharp single lines shown by

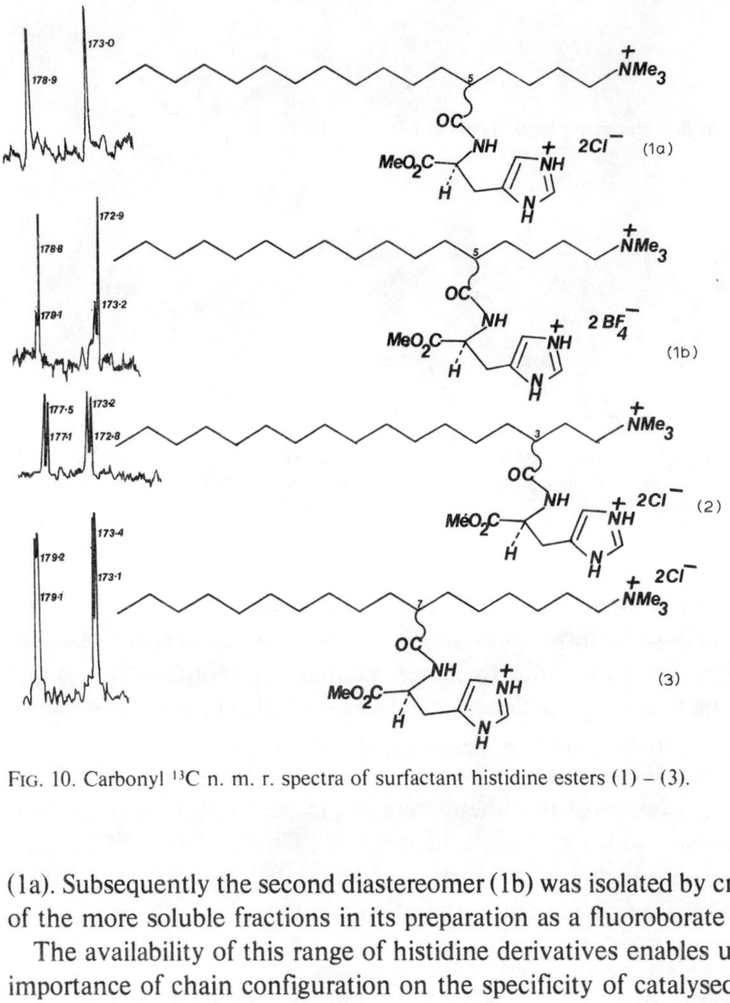

FIG. 10. Carbonyl ^{13}C n. m. r. spectra of surfactant histidine esters (1) – (3).

(1a). Subsequently the second diastereomer (1b) was isolated by crystallization of the more soluble fractions in its preparation as a fluoroborate salt.

The availability of this range of histidine derivatives enables us to test the importance of chain configuration on the specificity of catalysed hydrolysis. We interpret the results (Fig. 11) to suggest that only one diastereomer of (1) – (3) is a stereoselective catalyst, and therefore chain configuration is all important. A rational synthesis of the R-carboxylic acid precursor to (1) is in progress.

The important factors leading to asymmetric catalysis can be summarized as follows:

(*a*) in the reaction of (1a) with *N*-acetylphenylalanine *p*-nitrophenyl ester, enantiomeric recognition is a transition-state phenomenon (see Fig. 12) and

	S, k_{rel}	k_S/k_R
3-Histidine (mixed diastereomers)	0.92	1.50
5-Histidine:		
isomer A	1.00	3.10
isomer B	0.21	1.04
7-Histidine (mixed diastereomers)	0.51	1.67

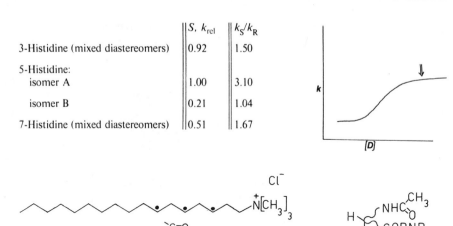

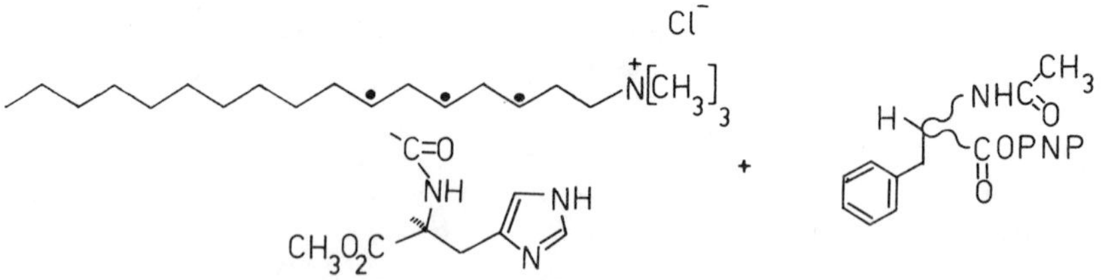

FIG. 11. Stereoselectivity in hydrolysis by chain-histidine surfactants (3-, 5- or 7-substituted): PNP, *p*-nitrophenyl; arrow indicates surfactant concentration used in these experiments.

no differentiation in binding to the micelle is observed;

(*b*) of substrates so far tested, only amino acid derivatives are effective, a fact which suggests that intra-amide hydrogen bonding is important;

(*c*) the configuration of the carboxy side-chain is crucial; the diastereomer which reacts the more rapidly is also the more selective.

Taken together, these factors allow us to construct an empirical model. Since all other recent examples of asymmetric catalysis in ester hydrolysis rely on the construction of an 'active site' analogue (see, *e.g.*, Chao & Cram 1976; Flohr *et al.* 1976) we believe that it is important to define the precise mechanism in this one case where no such pre-organization is imposed.

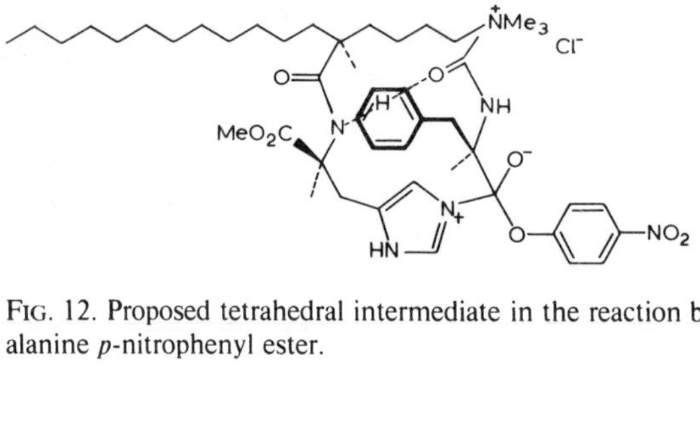

FIG. 12. Proposed tetrahedral intermediate in the reaction between (1a) and (*R*)-*N*-acetylphenylalanine *p*-nitrophenyl ester.

MECHANISMS OF ACYL TRANSFER IN MICELLES

The mechanisms of micelle-promoted reactions have usually been established by kinetic methods, using isotope effects, substrate variation and changes in electronic spectra as probes. We find that Fourier-transform ^{1}H n.m.r. spectroscopy at 270 MHz enables us to obtain satisfactory spectra from millimolar solutions in a few seconds' acquisition time. This enables the course of micellar reactions in water to be followed directly, and I shall describe three typically informative experiments. In the first, the histidine compound (1a) was treated in buffered ^2H_2O (pD 7.1) with p-nitrophenyl acetate at 313 K. A rapid conversion of starting ester into acylhistidine was observed, monitored by changes in the aromatic region, acyl methyl and, more surprisingly, the carbomethoxy regions of the spectrum. This was followed by slow hydrolysis of the resulting acylhistidine at a rate essentially independent of its concentration, the product being a mixture of acetyl phosphate (from the buffer) and acetate

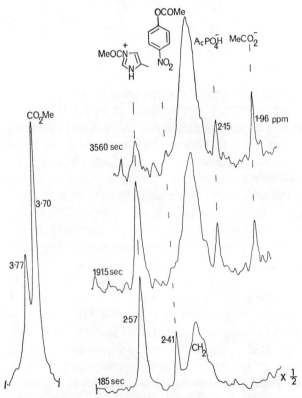

FIG. 13. ^{13}C n. m. r. indications of acetylation and deacetylation of 5-histidine compound (313 K, pD 7.1, 5 mmol/l).

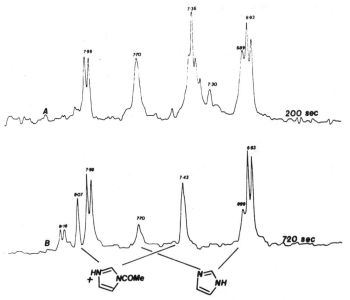

Fig. 14. N. m. r. spectrum of 5-histidine compound (313 K, pD 7.1, 5 mmol/l) plus (A) *N*-acetyl-phenylalanine *p*-nitrophenyl ester and (B) *p*-nitrophenyl acetate.

ion. Fig. 13 shows the spectroscopic changes which occur between 3 and 1.8 p.p.m. during the course of hydrolysis. Intermediacy of an acylhistidine is clearly apparent, and may also be seen by monitoring changes in the aromatic region of the n.m.r. spectrum (Fig. 14). No such intermediate is seen, to our surprise, during similar catalysed hydrolysis of *N*-acetylphenylalanine *p*-nitrophenyl ester, as the histidine residues remain unchanged throughout. Since the kinetic isotope effect, and pH rate-profile are similar for the amino-acid ester and for simple acid esters (see Fig. 9) there is unlikely to be a radical change of mechanism. It is more probable that the initially formed acylhistidine (4) is rapidly degraded *via* a (histidine-catalysed?) cyclic mechanism leading to the formation of oxazoline (5) and this idea is being tested.

A second set of experiments is related to attempts by ourselves and others (summarized in Kunitake & Okahata 1976; Kunitake *et al.* 1976; see also Moss *et al.* 1977) to demonstrate bifunctional catalysis in micelles. The method has been to seek evidence of participation of histidine in acyl transfer to the hydroxamic acid (6a) at a pH where both surfactant-bound functions are un-ionized (Fig. 15). Attempts to demonstrate this by kinetics are equivocal and we sought more definite information by n.m.r. spectroscopy. It was immediately apparent that (6a) alone, or its analogue (7a), is rapidly acylated by *p*-nitrophenyl acetate. In a mixture of (1a) and (6a) or (7a) only the acylhydroxamic acid was observed

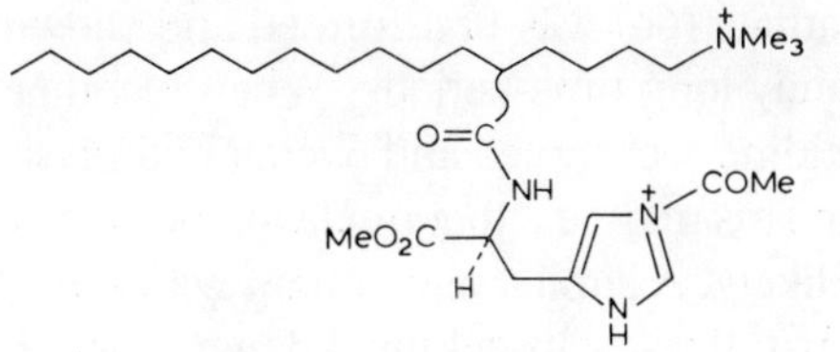

(**4**)

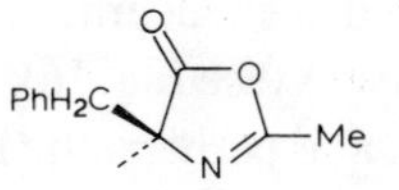

(**5**)

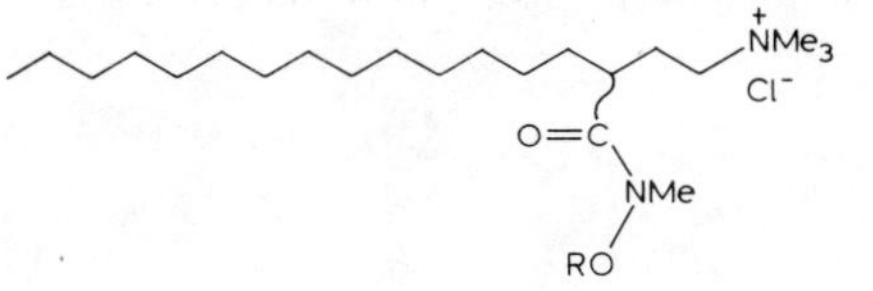

(**6**) *a*; R=H
b; R=$COCH_3$

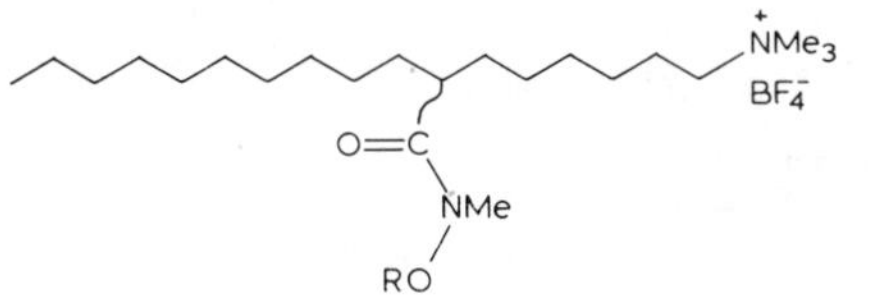

(**7**) *a*; R=H
b; R=$COCH_3$

after addition of *p*-nitrophenyl acetate, appearing to lend weight to the possibility of partial reaction *via* the mechanism of Fig. 15. This interpretation was effectively scotched by first preparing the acetyl-histidine *in situ,* adding (6a) and recording the n. m. r. spectrum immediately. The half-life of acyl-transfer is less than 20 s. A similar result was obtained using (7a) in place of (6a) although here reaction is marginally slower.

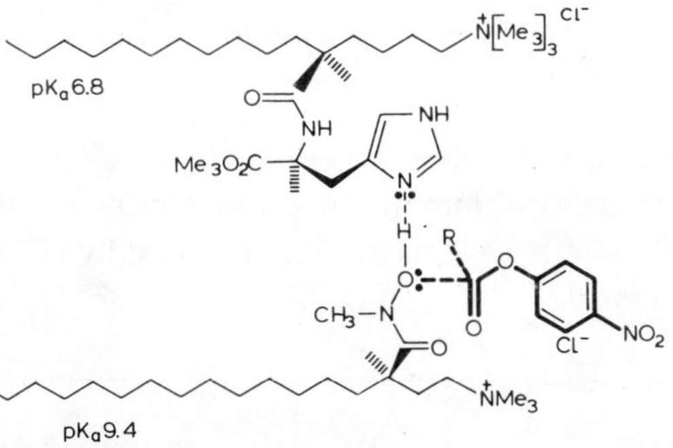

FIG. 15. Bifunctional micellar catalysis.

In the third experiment acylhydroxamate (6b) was first formed and shown to be stable to hydrolysis over moderately long time-periods. When histidine (1) is added, a small amount of acylhistidine is observed and overall hydrolysis of (6b) ensues (see Fig. 16). The mechanisms may be either nucleophilic or general-base, and perhaps the latter is less likely. A similar experiment with the 7-acylhydroxamate (7b) shows that its histidine-catalysed breakdown is much slower, perhaps partly owing to the more hydrophobic environment of the acyl-group in the former.

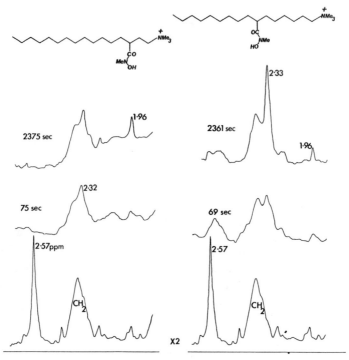

FIG. 16. Breakdown of hydroxamate catalysed by histidine (1); acyl transfer was followed by ^{13}C n. m. r. spectroscopy (pD 7.1, 5 mmol/1, 313 K).

It seems, therefore that bifunctional catalysis in micellar catalysis is not readily attainable. Nevertheless, the rapidity of intramicellar acyl transfer and the possibility of general-base catalysis in hydrolysis of (6b) promoted by (1) are useful pointers to future experiments.

CATALYSIS BY CATIONIC MICELLES: A SUMMARY

Micellar catalysis has been observed in nucleophilic addition and substitution of both saturated and unsaturated systems. In carbonyl chemistry, two of the three important mechanisms have been demonstrated.

(a) Direct attack of nucleophile at carbonyl carbon atom

Numerous examples now exist of micellar catalysis in attack of neutral or anionic nucleophiles on acyl derivatives. For example, the 'true' second-order rate constant for the reactions of imidazoles with p-nitrophenyl esters is lower in micelles than in bulk solution, although this is compensated by the advantage of approximation of reactants at the micellar surface. In mildly basic media, the effectiveness of catalysis increases, owing to reaction of the imidazolide anion. The acidity of micelle-bound imidazole is enhanced by several pK_a units because of a favourable electrostatic effect exerted by the cationic head-group on its conjugate anion.

(b) Proton transfer

Reports of the activation of bases by cationic micelles are rare, but increasing. The E1cB mechanism observed in hydrolysis of esters bearing an acidic α-hydrogen atom is an example (Tagaki et al. 1976). Thus hydrolysis of p-nitrophenyl benzyloxyacetate is feebly catalysed but that of p-nitrophenyl benzylthioacetate is catalysed dramatically. The pH–rate profile of reaction demonstrates a rate-determining deprotonation and, presumably, a (phenylthio)-ketene intermediate. Two factors will be important here. First, hydroxide ion at the micellar surface will be desolvated, and thus reactions will be characterized by a lower activation energy. Secondly, a delocalized carbanion will be better stabilized in the microenvironment of the micellar head-group.

There are few well documented examples of general-base catalysis, the third important reaction pathway of acyl-transfer reactions. This contrasts with enzyme studies, where, for example, synchronous proton transfers are an integral feature of the mechanism of action of serine proteinases. Kirby & Lloyd (1976) and Jencks (1975) have examined the evidence from model systems and concluded that intramolecular nucleophilic catalysis may provide rate-enhancements of up to 10^{15} mol/l but intramolecular general-base catalysis is rarely associated with a catalytic constant of more than 20 mol/l. This is because nucleophilic attack on an acyl centre follows a rigidly defined geometry, whereas the position of a proton in proton-transfer step is much less constrained. Consequently intramolecularity (or intermolecular approximation in the case of micelles) is much less of an advantage. General base-catalysed reactions are even less felicitous in micelles for those cases in which the base activates water towards the substrate, since the micellar surface is a water-poor environment.

HETEROGENEOUS MICELLE-LIKE SURFACES

The major problem inhibiting synthetic use of micellar catalysis is one of scale. The upper concentration limit of substrate is around 10^{-2} mol/l, with surfactant in excess to avoid swamping of the micelle and consequent inhibition. Anyone who has had the misfortune to have to extract an organic product from concentrated surfactant solution will avoid repeating the procedure.

We therefore looked for ways in which we could modify the surface of a polymer to resemble that of a cationic micelle. Two starting materials were used; one a 2% cross-linked microporous polystyrene (Merrifield resin), the second a macrorecticular polystyrene (Rohm and Haas XE-305), with 90.0 nm average pore-size. Both of these were chloromethylated to 100% substitution of available sites, and then treated as shown in Fig. 17. The chain linkage was established by trial and error; in earlier work (Brown & Jenkins 1976) a carboxylate ester was prepared which was labile in strong base.

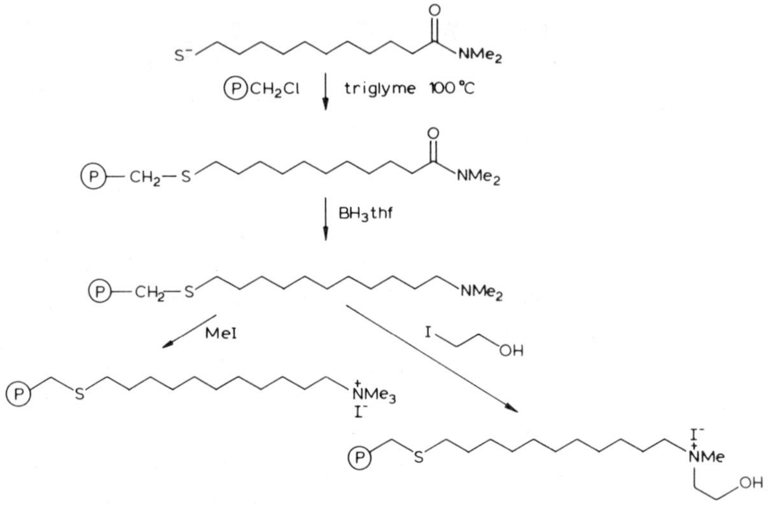

FIG. 17. Synthesis of micelle-like polymers.

The resulting polymers were free-flowing beads or powders with demonstrably amphiphilic properties. They swell somewhat in water, and counter-ion exchange is rapid. We hoped that the polymer–water interface would resemble the surface of a cationic micelle and catalyse similar reactions without the limitations of scale. The resins proved effective in conventional phase-transfer reactions, for example that between octyl bromide and aqueous sodium cyanide at 90 °C. We made a more quantitative assessment of their catalytic activ-

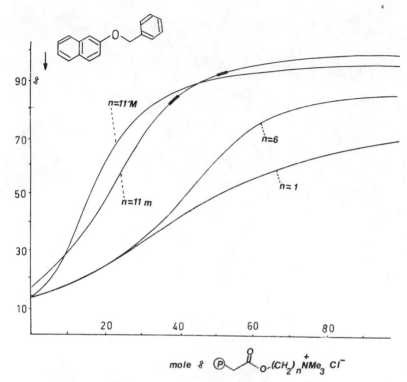

FIG. 18. Resin-promoted benzylation of 2-naphthoxide ion by benzyl bromide with P-CH$_2$·CO·O·[CH$_2$]$_n$·N$^+$Me$_3$ Cl$^-$.

ities by studying the alkylation of 2-naphthoxide ion by benzyl bromide (Figs. 18 and 19). This alkylates carbon in a protic environment but, in a polar aprotic environment where the oxygen is desolvated, alkylation of oxygen is the dominant process.

We were able to demonstrate that the resins catalyse alkylation in water and redirect the target from carbon to oxygen. In the course of measuring the binding constants for 2-naphthoxide ion on various resins, we noted that greater than stoichiometric amounts of the ion (based on effective molarity of quaternary ammonium groups) were absorbed, suggesting oligomeric 'stacks' of naphthoxide at the resin binding-site. In fact, concentrated aqueous solutions of sodium naphthoxide show a similar effect; we observed substantial changes in the chemical shift of the 6- and 7- protons in the ^{1}H n. m. r. spectrum above 0.5 mol/l (see Fig. 19).

There is further evidence that micelle-related resins can activate aqueous

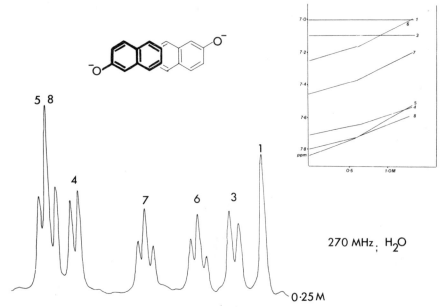

FIG. 19. ^{1}H n.m.r. spectrum of 2-naphthoxide ion at 270 MHz at 0.25 mol/l in H_2O. The inset shows the change in chemical shift of the protons (in p.p.m.) with increasing concentration.

hydroxide ion. This has been studied by measuring the rates of deuterium exchange in simple ketones. The most interesting aspect is the prospect of selective enolate-ion generation from αβ-unsaturated ketones, without any apparent catalysis of the competing β-hydration (Fig. 20).

Although all these reactions were designed to test mechanistic points, there is no barrier in principle to the use of micelle-related resins in preparative chemistry. There is the advantage over conventional phase-transfer catalysis that no solvent is needed; its role is fulfilled by the polymer surface.

ANALOGOUS WORK

I have described some attempts to effect organic reactions at surfaces and interfaces. This is one of several approaches to the problem of 'non-statistical recognition' in reacting systems, and three examples from recent work provide a conclusion and offer a pointer to future developments. The selective orientation of reactants in the solid-state is exemplified by recent work from the Weizmann school by Lahav *et al.* (1976) who resolved partially asymmetric anthroate esters of 1-phenylethanol by chiroselective photodimerization. Liquid

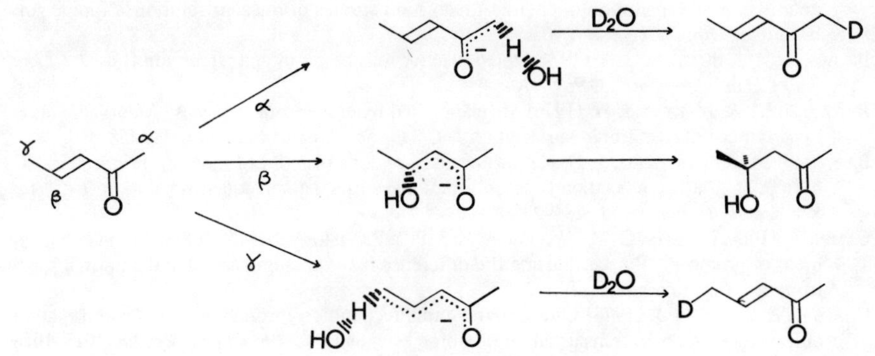

$t_{1/2}/s$	α	β	γ
NaOD	150	1800	»10^4
0·02M (P)C_{11},OD^-	50	5000	800
(P)C_{11} … O^-	50	5000	850

FIG. 20. Resin-promoted base-catalysed hydrogen-isotope exchange in pent-3-en-2-one.

crystals likewise provide a partly ordered environment, and there is one example (Saeva *et al.* 1975) of asymmetric reactions in a cholesteric mesophase. Finally, and again through work from the Weizmann school (Friedman *et al.* 1975), there is the selective oxidation of deoxycholic acid by photolysis of di-t-butyl peroxide oriented within its clathrate complex. No doubt many further examples of related results will appear in the near future.

ACKNOWLEDGEMENTS

I thank Professor Clifford A. Bunton for introducing me to micellar catalysis during a pleasant sabbatical leave in Santa Barbara, and colleagues and co-workers for the results described herein, particularly Drs John D. Schofield and Jesse L. Lynn. Professor Jerry A. Jenkins developed the micelle-related polymers during two successive summer visits to Oxford.

References

ANIANSSON, E. A. G., WALL, S. N., ALLINGTEN, M., HOFFMANN, H., KIELMANN, I., ULBRICHT, W., ZANA, R., LANG, J. & TONDRE, C. (1976) Theory of the kinetics of micellar equilibria and quantitative interpretation of chemical relaxation studies of micellar solutions of ionic surfactants. *J. Phys. Chem. 80*, 905

BROWN, J. M. & BUNTON, C. A. (1974) Stereoselective micelle-promoted ester hydrolysis. *J. Chem. Soc. Chem. Commun. 23*, 969–970

BROWN, J. M. & JENKINS, J. A. (1976) Micelle-related heterogeneous catalysis. Anion-activation by polymer-linked cationic surfactants. *J. Chem. Soc. Chem. Commun. 12*, 458–459

BROWN, J. M. & SCHOFIELD, J. D. (1975) Localised regions of reduced mobility in micelles; ^{13}C n. m. r. spin-lattice relaxation times of functional surfactants in aqueous solution. *J. Chem. Soc. Chem. Commun. 11*, 234–235

CAMPBELL, I. D., DOBSON, C. M., WILLIAMS, R. J. P. & ZAVIER, A. V. (1973) Resolution enhancement of proton PMR spectra using the difference between a broadened and a normal spectrum. *J. Mag. Resonance 11*, 172

CHAO, U. & CRAM, D. J. (1976) Catalysis and chiral recognition through designed complexation of transition states in transacylation of amino ester salts. *J. Am. Chem. Soc. 98*, 1015–1016

FENDLER, J. H. & FENDLER, E. J. (1975) *Catalysis in Micellar and Macromolecular Systems*, Academic Press, New York

FLOHR, K., PATON, R. M. & KAISER, E. T. (1976) Studies of the interaction of spin-labeled substrate with chymotrypsin and with cycloamyloses. *J. Am. Chem. Soc. 98*, 1209–1218

FRIEDMAN, N., LAHAV, M., LEISEROWITZ, L., POPOVITZ-BIRO, R., TANG, C.-P. & ZARETZKII, Z. (1975) Reactions in inclusion molecular complexes. A one-step regiospecific and stereospecific hydroxylation of deoxycholic acid. *J. Chem. Soc. Chem. Commun. 21*, 864–865

HOPF, F. R. & WHITTEN, D. W. (1976) Photochemical reactions in organized monolayer assemblies. 5. Photochemical and thermal reactions of reactive intermediates formed by ligand photoejection in ruthenium porphyrins. *J. Am. Chem. Soc. 98*, 7422–7424

ISRAELACHVILI, J. N., MITCHELL, D. J. & NINHAM, B. W. (1976). Theory of self-assembly of hydrocarbon amphiphiles into micelles and bilayers. *J. Chem. Soc. Faraday II 72*, 1525

JENCKS, W. P. (1975) Binding energy, specificity and enzymic catalysis– the circe effect. *Adv. Enzymol. 43*, 219–410

KIRBY, A. J. & LLOYD, G. J. (1976) Structure and efficiency in intramolecular and enzymic catalyses: intramolecular general base catalysed hydrolysis. Hydrolysis of monoaryl malonates. *J. Chem. Soc. Perkin Trans. II 14*, 1753–1761

KUNITAKE, T. & OKAHATA, Y. (1976) Multifunctional hydrolytic catalyses. 7. Cooperative catalysis of the hydrolysis of phenyl esters by a copolymer of N-methylacrylohydroxamic acid and 4-vinylimidazole. *J. Am. Chem. Soc. 98*, 7793–7799

KUNITAKE, T., OKAHATA, Y. & SAKAMOTO, T. (1976) Multifunctional hydrolytic catalyses. 8. Remarkable acceleration of hydrolysis of p-nitrophenyl acetate by micellar bifunctional catalysts. *J. Am. Chem. Soc. 98*, 7799–7806

LAHAV, M., LAUB, F., GATI, E., LEISEROWITZ, L. & LODMER, Z. (1976) A new method of enantiomeric purification via a topochemical photodimerization reaction. Application to three l-aryl ethanols. *J. Am. Chem. Soc. 98*, 1620–1622

MOSS, R. A., NAHAS, R. C. & RAMASWAMI, S. (1977) Sequential bifunctional micellar catalysis. *J. Am. Chem. Soc. 99*, 627–629

OAKENFULL, D. (1973) Effects of hydrophobic interaction of the kinetics of the reactions of long chain alkylamines with long chain carboxylic esters of 4-nitrophenol. *J. Chem. Soc. Perkin Trans. II 7*, 1006–1012

RIDEAL, E. K. (1945) Reactions in monolayers. *J. Chem. Soc.* 423

RIDEAL, E. K. & DAVIES, J. T. (1961) *Interfacial Phenomena*, Academic Press, London & New York

SAEVA, F. D., SHARPE, P. E. & OLIN, G. R. (1975) Asymmetric synthesis in a cholesteric liquid crystal solvent. *J. Am. Chem. Soc. 97*, 204–205

SPRINTSCHNIK, G., SPRINTSCHNIK, H. W., KIRSH, P. P. & WHITTEN, D. W. (1976) Photochemical cleavage of water: a system for solar energy conversion using monolayer-bound transition metal complexes. *J. Am. Chem. Soc. 98*, 2337–2338

TAGAKI, W., KOBAYASHI, S., KURIHARA, K., KURASHIMA, A., YOSHIDA, Y. & YUMIHIKO, Y. (1976)

Effect of cationic micelles on the $E1cB$ mechanism of the hydrolysis of substituted p-nitrophenyl acetate. *J. Chem. Soc. Chem. Commun. 21*, 843

VALENTY, S. J. & GAINES, G. L. JR. (1977) Preparation and properties of monolayer films of surfactant ester derivatives of tris (2,2'-bipyridine)ruthenium(II)$^{2+}$. *J. Am. Chem. Soc. 99*, 1285-1287

Discussion

Baldwin: What evidence is there for the common picture of a micelle as a spherical object with all the tails pointing inwards?

Brown: Spheroidal structures of low dispersisity are definitely seen in solution. In other words, dimensions are related simply to the length of the alkyl chain; light-scattering, particularly, and other physicochemical experiments demonstrate that (Fendler & Fendler 1975).

Baldwin: What happens at the middle of the micelle where all the tails meet? Would a bilayer structure be more appropriate?

Breslow: Not all the chains go into the centre according to our recent (unpublished) work; many fill the intervening spaces.

Brown: We have to recognize that the surface is rough and that the structure is irregular. The small micelle can be modelled mathematically with those constraints and without having infinite density at the centre.

Baker: This surface irregularity must be important for without this ruffling and curling back of the tails one could regard micellar catalysis as little more than an alternative to catalysis by any other supported catalyst. The body of the micellar catalysis also provides an environment for favourable reaction.

Brown: The environment of the catalyst differs greatly from that of a similar catalytic species in isotropic solution.

Baker: In metal catalysis on a support material the catalyst is provided at the surface and the support has no part in the reaction.

Brown: But in solid-state catalysis there are distinct effects which do not have any counterpart in dilute solution. For example, the distribution patterns in olefin hydroformylation are different with polymer supported organometallic catalysts from those with their counterparts in solution (Pittman & Hanes 1976).

McCapra: These micellar reactions do not become interesting until the solvent effect has been eliminated. What would be the effect of using a dipolar aprotic solvent such as Me_2SO for your last reaction?

Brown: The molar concentration of Me_2SO needed to divert the alkylation of naphthoxide from carbon to oxygen is about twice as high as the effective molar concentration of the polymeric quaternary-ammonium substituted resin needed to effect the same degree of diversion.

Baldwin: In the hydrolysis of the *p*-nitrophenyl esters in which there is selection between the two enantiomers, what happens if you do away with the long hydrocarbon chain—the micellar part of the reaction—just take histidine or an acylhistidine methyl ester? Does what you observe have anything to do with the micellar structure?

Brown: What about the case where we change the configuration at the chain and discrimination disappears?

Baldwin: That would be expected if there were a 2-butyl centre there. You are still making a chiral base. There should be selection with a simple histidine derivative.

Brown: Nobody has been able to show this, but Sheehan *et al.* (1966) found that some small synthetic histidine-containing peptides can enantioselectively hydrolyse *N*-acetylphenylalanine methyl ester.

Baldwin: So, according to you, *N*-acetylhistidine methyl ester should not differentiate between the two enantiomers of the *p*-nitrophenyl esters on hydrolysis?

Brown: Yes; but we have not done the experiment.

Baldwin: I should have thought it would.

Brown: If, as seems to be the case for a range of esters with and without amino-acid residues, some hydrogen-bonding component is necessary for selectivity, it is clearly an advantage to do the reaction in a water-poor environment.

Battersby: Do the relaxation times for the protons at the polar end and at the branch support the ^{13}C n. m. r. results?

Brown: I don't know. It is more difficult to study ^{1}H than ^{13}C n. m. r. relaxation times; many more constraints are imposed, including deoxygenation of samples and an accurately spherical sample cavity.

Dewar: A related factor which tends to be overlooked is that polar reactions in non-polar solvents can show micelle-like behaviour due to aggregation. A good example of this is the reaction (1) of *N*-bromoacetanilide and anisole catalysed by acid in CCl$_4$. At high concentrations of anisole, the reaction was first-

$$\text{PhNBrAc} + \text{PhOMe} \xrightarrow{\text{AcOH}} \text{PhNHAc} + \text{Br-C}_6\text{H}_4\text{-OMe} \tag{1}$$

order in *N*-bromoacetanilide and at high concentrations of the acetanilide it was first-order in anisole but in each case zero-order in the other component. (It is unusual to find a bromination zero-order in the brominating agent.) This behaviour is due to the clustering together of the molecules. With an excess of anisole, every cluster containing *N*-bromoacetanilide will contain at least one anisole molecule to react with. More anisole, therefore, will not alter the rate, so the reaction will be independent of the concentration of anisole. Equally, if there is a large excess of the acetanilide, each molecule of anisole will be clustered with at least one of the brominating agent. Again, the lack of effect of the excess means that the rate will be independent of concentration. The lack of homogeneity in these conditions casts doubt on any conclusions drawn from kinetic results.

Brown: That would be an unconventional explanation for a species such as anisole. Many examples of reaction in non-polar aggregates are known (see Fendler 1976). Long-chain hydrocarbons carrying a polar terminus form aggregates in non-polar solvents containing a trace of water which are basically a microscopic pool of water surrounded by a hydrophobic core. These can effectively catalyse a range of ionic organic reactions. But anisole should not aggregate in any solvent.

Dewar: I am talking about small clusters of polar molecules, not large aggregates. All the acetic acid will be hydrogen-bonded to anisole or the acetanilide and more molecules may stick to them to form a small cluster. The solution is not truly homogeneous. Only when all the reagents are gathered in one cluster will they react.

Brown: There is a great tendency for organic molecules to aggregate in solution; most of the time this goes unrecognized.

Kenner: I noticed that the polymer was selectively N-alkylated. If the thioether had been attacked, chains could have been degraded and lost.

Brown: Once the selective N-methylation is effected, the sulphide seems to be protected, but we had to choose conditions for reaction of the amino-sulphide with methyl iodide extremely carefully. Since the surface of our polymers is fluid we can assay reactions by ^{13}C n.m.r. spectroscopy in $[^2H_6]Me_2SO$ and get moderately sharp resonances from nuclei close to the surface.

References

FENDLER, J. H. (1976) Interactions and reactions in reversed micellar systems. *Acc. Chem. Res. 9,* 153–160.

FENDLER, J. H. & FENDLER, E. H. (1975) *Catalysis in Micellar and Macromolecular Systems,* ch. 2, Academic Press, New York

PITTMAN, C. U. JR. & HANES, R. M. (1976) Unusual selectivities in hydroformylations catalyzed by polymer-attached $(PPh_3)_3RhH(CO)$. *J. Am. Chem. Soc. 98,* 5402–5405

SHEEHAN, J. C., BENNETT, G. B. & SCHNEIDER, J. A. (1966). *J. Am. Chem. Soc. 88,* 3455

Studies on enzyme models and on the enzyme carboxypeptidase A

RONALD BRESLOW

Department of Chemistry, Columbia University, New York

Abstract The enzyme carboxypeptidase A has been extensively studied and is thus a good candidate for chemical models and imitation. A model system was constructed which combined two of the three catalytic functional groups of the enzyme together with the appropriate substrate group, and cooperative effective catalysis was demonstrated. However, the mechanism which the model system used in enzyme-like conditions was unexpected, and this led to a study of the enzyme itself.

Carboxypeptidase A was studied by examination of oxygen-18 exchange reactions and the ability of the enzyme to substitute other lytic agents, such as methanol, for the water which is normally used. The result of these studies is a proposed mechanism of action of this enzyme which accommodates the large amount of information already available, and which indicates that this enzyme uses a mechanism similar to that found in the model system.

Approaches have also been made to an artificial enzyme which could hydrolyse amides. Cyclodextrin has been functionalized with two imidazole groups placed on opposite sides of the cavity. This material catalyses the hydrolysis of an amide, which binds to the cyclodextrin cavity and then is hydrolysed with the assistance of one imidazole in its basic form and the other one in its protonated form, acting together in a way reminiscent of the cooperation of some of the functional groups of hydrolytic enzymes.

The enzyme carboxypeptidase A (EC 3.4.12.2) catalyses the hydrolysis of peptides at the COOH-terminal residue. Extensive previous work, and particularly the X-ray structures (Lipscomb *et al.* 1968), have established the outline of a mechanism for its action. Lipscomb studied the structure of carboxypeptidase A with a glycyltyrosine bound to the enzyme. Glycyltyrosine is actually an extremely poor substrate for the enzyme, which is why it is possible to study the complex, but it is hoped that the structure of the peptide–enzyme complex does not differ grossly from that of other peptides which serve as good substrates. The essential features of the structure are shown in Fig. 1; the terminal car-

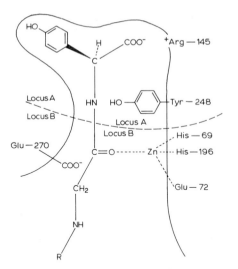

FIG. 1. Active site of carboxypeptidase, showing bound substrate (based on crystallographic structure): GlyTyr, R = H; substrate, R = R'CO etc.

boxylate group of the substrate is bound to an arginine of the enzyme, the tyrosine side chain is in a hydrophobic pocket of the enzyme, and the scissile carbonyl group is coordinated through its oxygen atom to the zinc ion of the enzyme. The other two important functional groups are a carboxylate group of the enzymic glutamic acid-270 and a phenolic hydroxy group of the enzymic tyrosine-248. The enzyme thus incorporates a basic carboxylate anion, an acidic metal cation and the acidic phenol in a structure so that these groups can cooperate in the hydrolysis of the peptide.

In the structure of the complex before hydrolysis begins the glutamic acid carboxylate anion is not very close to the carbonyl carbon to be attacked, but the reaction path could bring it closer to that carbon atom. In general two classes of mechanisms have been considered (Lipscomb 1974) by which the carboxylate could assist the cleavage of this peptide. In one the carboxylate directly attacks the carbonyl group and, after other appropriate changes, the glycyl residue (if this were an active substrate) would be transferred to the glutamate carboxyl group forming an intermediate acyl-enzyme, which is actually an anhydride. Hydrolysis of this intermediate anhydride would complete the overall process and could be catalysed by the zinc of the enzyme. We have reported elsewhere (Breslow *et al.* 1975) a model system for such zinc-catalysed hydrolysis of anhydride which indicates that the proposed overall process is reasonable and could occur on the enzymic time scale.

In an alternative mechanism (Lipscomb 1974) the glutamate carboxylate acts as a general base rather than a nucleophile and delivers a water molecule to the carbonyl group of the substrate. By this mechanism there is no acyl-enzyme intermediate, and the overall hydrolysis is essentially a one-step process, in the sense that no discrete intermediate is formed in which one of the parts of the substrate has already been lost.

We have been studying several model systems for the action of this enzyme, including one (Breslow & McAllister 1971) in which a zinc ion and a carboxylate ion cooperate in internal catalysis. Recently we took up the study of a model system for the cooperative effect of a carboxylate and a phenolic hydroxy group in the cleavage of an amide group. This model system (Breslow & McClure 1976) makes use of two of the three important functional groups of the enzyme and holds them in a well defined position relative to the substrate grouping, the amide linkage to be cleaved. The particular system studied is a maleamic acid derivative in which the nitrogen carries a phenolic group (Fig. 2). Maleamic acids are well known and undergo rapid cleavage of the amide

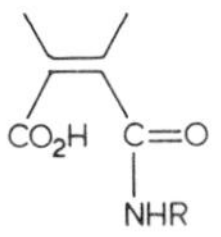

FIG. 2. The model maleamic acid derivative used in the cleavage studies. For R groups, see Table 1.

linkage in an acidic medium (cf. Kirby & Fersht 1971). That is, a carboxylate group can readily attack the amide carbonyl group in an internal reaction and, with an extra proton supplied from the acidic medium, the resulting tetrahedral intermediate decomposes to cleave the carbon-nitrogen bond and produce a maleic anhydride. In our model system we at first expected that the phenol might act as a general acid to replace the proton supplied from an acidic medium. If this were the case this same mechanism could operate in neutral solution, so bringing the pH near to that at which the enzyme functions. However, the actual findings with this model system were much more interesting than this simple idea suggests (see Table 1). In an aqueous solution the phenolic hydroxy group was not catalytic at any pH. That is, the cleavage of the maleamic acid needed a proton from the medium, and the phenolic hydroxy group was not able to substitute for this proton. However, the interior of an enzyme is not really aqueous. Instead, one of the important functions of the protein is to exclude water from the active site, at least to some extent, so as to make the reaction region much more like a highly polar organic medium such as dimethylformamide or dimethyl sulphoxide.

TABLE 1
Effect of substituent R in the maleamic acid derivative shown in Fig. 2 on rate of cleavage. All rates were the same in H_2O through pH 4–9. The solvent used was CH_3CN, 1.0 mol/1 in H_2O.

R	Rate ($10^4 k/s$) with HOAc/KOAc buffer		
	10:1	1:10	1:50
$C_6H_5CH_2-$	75.3	~0.1	0.0314
2-hydroxybenzyl	45.5	1.80	2.07
2-hydroxy-5-bromobenzyl	41.6	5.12	4.55
indanyl	28.8	~0.07	0.0238
hydroxyindanyl	20.1	0.866	0.959

This function of the enzyme in excluding water can be imitated in model systems simply by changing the solvent. We studied the behaviour of the maleamic acid derivatives in 1M-acetonitrile (in H_2O). In such a medium there is some problem about the definition and significance of pH. We solve this by defining the basicity or acidity of the medium in terms of the buffer ratio which we add. Thus with 10/1 acetic acid/potassium acetate buffer, we are dealing with a medium which is acidic. With a 1/50 ratio of acetic acid/potassium acetate, we are dealing with a buffer ratio which would correspond to neutrality if it were in water. We therefore characterize it as being essentially a neutral medium even in the acetonitrile–water system.

Strikingly, we find that in this medium the phenolic hydroxy group of our model is catalytic. It does not accelerate the reaction when the acidic buffer system is used, since apparently the proton is best supplied in that case from the medium. However, in the neutral system we now find that the maleamic acid carrying a phenolic hydroxy group is about 70 times as reactive as the one without. The substituent effects in acidic medium are reversed, for electronic reasons, so that the real catalysis by the phenolic hydroxy group is a little in excess of 100-fold.

The most obvious explanation of this finding would be that the phenolic hydroxy group can now act as a general acid to supply the proton needed for the conversion of maleamic acid into the maleamic anhydride. However, two facts argue against this. One is that the acidity of the phenol does not seem to be nearly as important as would be expected if it were acting as a general acid to replace a proton from the medium. Thus the *p*-bromo derivative is only a little more than twice as active, although the phenolic hydroxy group in this case should be about 10 times more acidic. This means that the Brönsted coefficient for catalysis is small, a surprising result for a process which ordinarily needs a full proton transferred from the medium.

The second argument against this simple picture is much more pursuasive: the anhydride is no longer formed in the reaction. Instead, the maleamic acid derivative bearing a phenolic hydroxy group undergoes a direct hydrolysis at neutrality to produce the maleic acid without the intervention of an anhydride intermediate. The anhydride would be readily detected spectroscopically and would have an appreciable lifetime in these conditions. When authentic dimethylmaleic anhydride is added to the medium, it can be readily detected, but it is not formed during the hydrolysis of the maleamic acid. This is also indicated by the fact that the authentic dimethylmaleic anhydride can be trapped by amines added to the system, to produce other maleamic acid derivatives, but the addition of amines to our phenolic maleamic acid model compound during the course of its hydrolysis leads to no trapping of an intermediate anhydride. Apparently the compound is hydrolysed by a different mechanism than the nucleophilic one; it must therefore be being cleaved by a general base catalysed delivery of water to the carboxamide group (controls exclude other possibilities, such as nucleophilic attack by the phenol group).

If the carboxylate anion of the maleamic acid is delivering water to the carboxamide, then we must still wonder about the function of the phenolic hydroxy group in facilitating the overall hydrolysis. The best evidence on this comes from a consideration of the fact that the mechanism change has to be understandable. If the general base mechanism is now preferred (whereas with a good protonating medium the nucleophilic mechanism is preferred instead), this must be because the general base mechanism does not require a subsequent protonation of the substrate groups as the anhydride mechanism did. Otherwise the nucleophilic mechanism, leading to an anhydride, would always be the preferred mechanism regardless of which group was used to supply the proton needed to finish off the reaction. Thus the phenolic hydroxy group must be serving some function other than simple protonation of an intermediate. This also follows from our observation that the acidity of this phenol was much less important than might have been expected based on the earlier considerations.

All this can be best understood in terms of a function of the phenolic hydroxy group in facilitating the interconversion of two different forms of the tetrahedral intermediate. In many hydrolyses of carboxamides it has been shown (cf. Jencks 1969) that the first-formed tetrahedral intermediate, in which the central carbon atom carries an OH group and a neutral nitrogen, must be converted by a proton transfer into an isomeric structure in which the nitrogen is protonated and the oxygen is not. The resulting alkoxide anion can then expel the protonated nitrogen, which on departure becomes a neutral amine.

In water medium the proton will be transferred between the hydroxy group and the nitrogen of the tetrahedral intermediate commonly by the solvent itself. However in the essentially non-aqueous medium we are using to imitate the interior of the enzyme, such a proton transfer will not be rapid and can be subject to catalysis. We believe that the phenolic hydroxy group is catalysing such a proton transfer, as shown in Fig. 3. It acts as a general acid to protonate

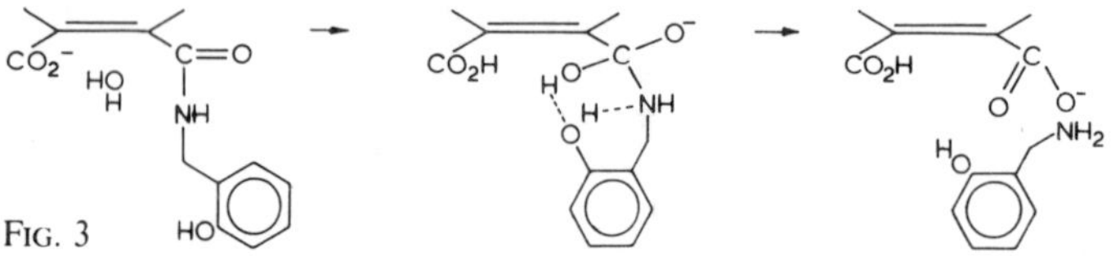

FIG. 3

the amino group, as was expected even for the nucleophilic anhydride mechanism. However the resulting phenoxide anion then also acts as a general base to remove the proton from the hydroxy group of the tetrahedral intermediate. The exact timing of these steps is not established, and they may well be concerted by the switching of a couple of bond lengths in some hydrogen bonds, but it is likely that protonation of the nitrogen is a little more advanced than proton recovery from the hydroxy group of the tetrahedral intermediate if we are to explain the fact that the acidity of the phenol has some effect on the rate, if not a large one.

Thus in our model system the carboxylate ion is serving the function of a general base, to deliver water to the carbonyl, while the phenolic hydroxy group is serving the function of transferring protons among the two heteroatoms in the resulting tetrahedral intermediate. On account of these findings we reexamined the data on carboxypeptidase A itself to see whether perhaps a related mechanism operates for the enzyme.

As we have recently described our studies on the enzyme carboxypeptidase A in some detail (Breslow & Wernick 1977) I shall not repeat the details here. The essential points are as follows: in considering whether the enzyme uses a general base or a nucleophilic mechanism, we realized that we could test this by studying ^{18}O exchange for, if the enzyme uses a nucleophilic mechanism to form an anhydride intermediate which then hydrolyses, then the water involved in the overall hydrolysis comes in only in the second step of the process. By contrast, a general base mechanism for cleavage of the peptide involves the hydrolytic water immediately. The distinction between these two is more easily seen if we consider that the enzyme must be able to catalyse any hydrolysis in both directions, that is it must be able to resynthesize the peptide from the two hydrolytic fragments. Of course the equilibrium constant for hydrolysis of the peptide is large so that the rate of resynthesis of a peptide is expected to be small compared to the rate of hydrolysis. However, enzymic hydrolysis rates are in general so large that we would still expect an appreciable rate of resynthesis, and this can be detected by appropriate ^{18}O studies.

In the key experiment benzoylglycine was labelled with ^{18}O in the terminal carboxy group, and we examined the ability of carboxypeptidase A to catalyse the exchange of this ^{18}O with the normal oxygen of the medium (water). We found that the enzyme could not catalyse such ^{18}O exchange unless a second component was added to the system. The most effective second component was phenylalanine, and this effect is not surprising. The enzyme must be able to catalyse the condensation of benzoylglycine with phenylalanine to produce the corresponding peptide, and such a condensation followed by enzyme-catalysed hydrolysis to the original components is of course a mechanism by which ^{18}O exchange can occur. Such a mechanism must operate, and such ^{18}O exchange then must be seen regardless of the mechanism of the hydrolysis.

However, if the enzyme uses a two-step sequence, with an anhydride intermediate, one might expect to observe ^{18}O exchange in the absence of a second component such as phenylalanine; that is, in the synthetic direction, one would expect by the nucleophilic mechanism that the enzyme would first condense with benzoylglycine to form an intermediate anhydride. This would then react with phenylalanine in a second step to produce the peptide. The point is that in the first step ^{18}O exchange should have occurred if the anhydride were formed and then hydrolysed again, even if phenylalanine had not been present to send the anhydride intermediate all the way back to peptide. It is known that if such an anhydride intermediate is formed it must be cleaved on the side away from the enzyme carboxy group, since in the course of normal hydrolysis with $H_2^{18}O$, ^{18}O is not incorporated into the enzyme (Nau & Riordan 1975). Thus by this mechanism we should have observed exchange of the carboxy oxygen atom of benzoylglycine even in the absence of phenylalanine.

There is a possibility that the phenylalanine plays a role other than that suggested by the simple interpretation of these experiments. For instance, the binding of phenylalanine to the enzyme may be needed to induce the correct folding of the protein so that it can carry out the rest of the catalytic reaction. On this basis then, one might still allow an anhydride intermediate in the reaction and simply need phenylalanine for inducing the correct shape of the enzyme. But this explanation seems most unlikely. As we have described (Breslow & Wernick 1976, 1977), there is an excellent parallel between the ability of various second components to act as co-catalysts for the ^{18}O exchange and their expected ability to condense with the benzoylglycine to form a peptide. We can roughly calculate the rate of ^{18}O exchange which must occur from peptide resynthesis, since the equilibrium constant for peptide hydrolysis is known and we also know the rate constant for enzymic cleavage of the peptides. The usual relationship between rate constant and equilibrium constant applies here as well, although the treatment is a little more complex; enzymes show kinetics involving binding and inhibition, and thus the full rate expression involves various binding constants. However using this rate expression we can calculate the expected rate of ^{18}O exchange in benzoylglycine in the presence of various second components such as phenylalanine, and in every case the calculated constant is close to that observed. This fact and related arguments strongly suggest that the enzyme does not use a two-step mechanism for cleavage of peptides, but instead uses a direct hydrolysis mechanism. Therefore in the reverse direction it needs the full resynthesis of a peptide before ^{18}O exchange can be observed in benzoylglycine.

Another study on the enzyme also ties in well with the mechanistic conclusions derived from our model system. We have examined the ability of the enzyme to substitute methanol for the hydrolytic water in the transition state (Breslow & Wernick 1976, 1977). This cannot be observed in the forward direction for peptide cleavage, since methanolysis of a peptide would be endothermic and thus not observable. However, using the argument that enzymes must be able to catalyse a reaction in both directions, we can ask whether the enzyme can condense a methyl ester, such as the methyl ester of benzoylglycine, with an amino acid such as phenylalanine to produce benzoylglycylphenylalanine and methanol. This would be the exothermic direction of reaction and, if methanol can be used in the transition state for peptide cleavage, it must be acceptable in both directions. Our studies on this reaction demonstrate that water is preferred to methanol in the transition state by a factor of at least 6000; in fact, we have no evidence that methanol can participate at all. A similar finding is made for other small species, such as ammonia and hydroxylamine (for details see Breslow & Wernick 1977).

The meaning of this finding is interesting, since for many other hydrolytic enzymes such as trypsin and chymotrypsin methanol is preferred to water. Methanol, a small molecule which generally can get into the active site of an enzyme, is a better nucleophile in simple organic chemistry as well, so the results with other enzymes are not surprising. The striking contrast in our case with the normal finding suggests that there is a special reason why this enzyme needs water. Among the various possible differences, perhaps the most obvious one is that both protons of water may be needed for the mechanism. If so, this then suggests an enzyme mechanism shown in Fig. 4.

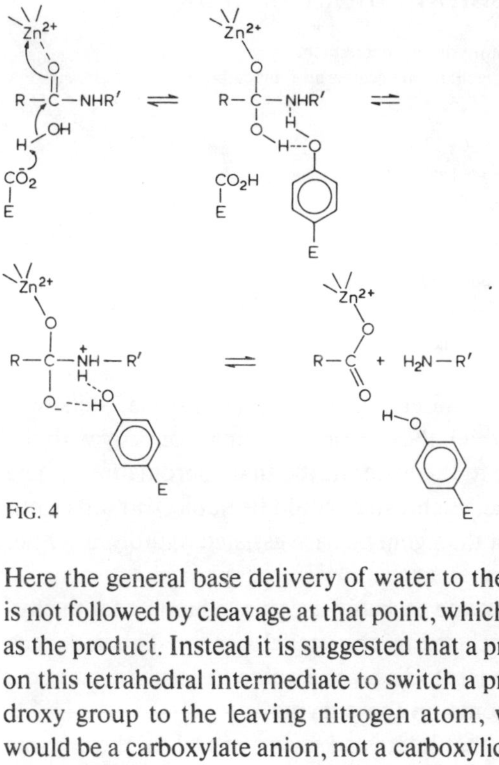

FIG. 4

Here the general base delivery of water to the carbonyl group of the substrate is not followed by cleavage at that point, which would produce a carboxylic acid as the product. Instead it is suggested that a proton transfer must be performed on this tetrahedral intermediate to switch a proton from the newly formed hydroxy group to the leaving nitrogen atom, with the result that the product would be a carboxylate anion, not a carboxylic acid. In such a mechanism both protons of the water were needed and used. Since the enzyme is using a phenolic hydroxy group, and since in our model system the phenolic hydroxy group performed exactly the function we have now described, we suggest that it does so in the enzyme as well. Thus the full mechanism we suggest involves the use of the tyrosine hydroxy group to facilitate proton transfer between the two heteroatoms of the tetrahedral intermediate.

This situation as described seemed fairly clear, but recently a striking finding has been reported by Makinen *et al.* (1976) who studied the carboxypeptidase A-catalysed hydrolysis of an ester, *p*-chlorocinnamoyl-β-phenyllactic acid.

This ester is cleaved by carboxypeptidase through a nucleophilic mechanism, in which an intermediate *p*-chlorocinnamoyl anhydride is formed.

However, as we have described (Breslow & Wernick 1977), the *p*-chlorocinnamoyl group is a curious one since the corresponding amide is a poor substrate for the enzyme. In any case several other pieces of evidence suggest that esters and amides may not be cleaved by analogous mechanisms. Instead we have suggested, as shown in Fig. 5, that although peptides are cleaved by the

---

HYPOTHESES CONSISTENT WITH ALL THE DATA

1. Peptide substrates use the direct, not anhydride, mechanism.
2. Ester substrates *can* use the anhydride mechanism because an ester carbonyl is not basic enough to bind well to Zn^{2+}.

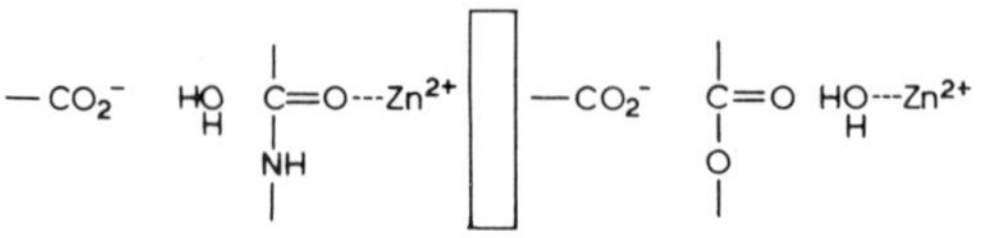

$$-CO_2^- \quad \overset{H}{\underset{}{HO}} \quad \overset{|}{\underset{NH}{C}}=O\cdots Zn^{2+} \qquad \qquad -CO_2^- \quad \overset{|}{\underset{O}{C}}=O \quad \overset{}{\underset{H}{HO}}\cdots Zn^{2+}$$

This explains the ability of Co^{3+}-carboxypeptidase to hydrolyse esters.

FIG. 5. Binding of esters and amides for cleavage.

general base mechanism as we have described above, at least some esters may bind in an alternative fashion in which they do not coordinate directly with the metal but instead are placed next to the water in the first coordination sphere of the metal. If this were the case, such esters could be subject to direct nucleophilic attack (see Fig. 5) rather than general base assisted hydrolysis. Other evidence supports this suggestion.

At present, it seems that the enzyme probably uses a mechanism for the hydrolysis of peptides similar to that used by the model system. This indicates a most important use of such model studies—as a guide to mechanism and a stimulus to mechanistic experiments on the enzyme.

Another kind of model system that interested us is a model for the enzyme which would catalyse the cleavage of a separate substrate. We have been working for many years on the use of cyclodextrin derivatives to produce such artificial enzymes (Breslow 1971). Although much of this work is not directly related to carboxypeptidase, one recent observation bears on the kind of acid-base mechanism we have outlined above for peptide cleavage. We prepared a derivative of β-cyclodextrin (cycloheptaamylose) with two imidazole groups on the primary carbon atoms, located on opposite sides of the cavity (Fig. 6), by treating Tabushi's bridged disulphonate ester (Tabushi *et al.* 1976) with imidazole.

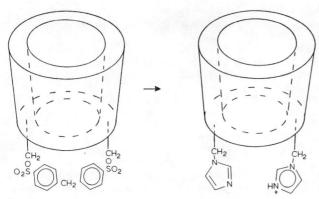

Fig. 6

The compound is of great interest as an enzyme model, particularly for ribonuclease in which two imidazoles are the two important functional groups of the enzyme. For my present purpose, I need only report that this compound serves as a hydrolytic catalyst for the cleavage of the reactive amide trifluoroacetyl-p-nitroanilide. This amide is cleaved in aqueous solution by binding to the cyclodextrin and subsequent cooperative catalysis by the two functional groups.

The pH–rate profile of this hydrolysis is a bell-shaped curve, with a maximum when one of the imidazole rings is protonated but the other one is not. This behaviour is highly reminiscent of that of ribonuclease, which shows a similar bell-shaped curve, and indicates that the two catalytic functions are operating respectively in their acidic and basic forms. Although we do not yet have a detailed mechanism for this hydrolysis analogous to that which we have suggested above for the other amide cleavages, we do have in this case the first bifunctional artificial enzyme which uses an acid-base mechanism to cleave an amide, as at least some of the natural amide-cleaving enzymes do.

References

Breslow, R. (1971) Studies on enzyme models. *Adv. Chem. Ser.* 100, 21–43

Breslow, R. & McAllister, C. (1971) Intramolecular bifunctional catalysis of ester hydrolysis by metal ion and carboxylate in a carboxypeptidase model. *J. Am. Chem. Soc.* 93, 7096–7097

Breslow, R. & McClure, D. (1976) Cooperative catalysis of the cleavage of an amide by carboxylate and phenolic groups in a carboxypeptidase model. *J. Am. Chem. Soc.* 98, 258–259

Breslow, R. & Wernick, D. (1976) On the mechanism of catalysis by carboxypeptidase A. *J. Am. Chem. Soc.* 98, 259–261

Breslow, R. & Wernick, D. (1977) A unified picture of the mechanisms of catalysis by carboxypeptidase A. *Proc. Natl. Acad. Sci. U.S.A.* 74, 1303–1307

Breslow, R., McClure, D. E., Brown, R. S. & Eisenach, J. (1975) Very fast zinc-catalyzed hydrolysis of an anhydride. A model for the rate and mechanism of carboxypeptidase A catalysis. *J. Am. Chem. Soc. 97*, 194–195

Jencks, W. (1969) *Catalysis in Chemistry and Enzymology*, pp. 487–496, McGraw-Hill, New York

Kirby, A. & Fersht, J. (1971) Intramolecular catalysis. *Progr. Bioorg. Chem. 1*, 1–82

Lipscomb, W. (1974) Relationship of the three dimensional structure of carboxypeptidase A to catalysis. *Tetrahedron 30*, 1725–1732

Lipscomb, W., Hartsuck, J., Reeke, G., Quiocho, F., Bethge, P., Ludwig, M., Steitz, T., Muirhead, H. & Coppola, J. (1968) The structure of carboxypeptidase A. VII. The 2.0 Å resolution studies of the enzyme and of its complex with glycyltyrosine, and mechanistic deductions. *Brookhaven Symp. Biol. 21*, 24–90

Makinen, M., Yamamura, K. & Kaiser, E. (1976) Mechanism of action of carboxypeptidase A in ester hydrolysis. *Proc. Natl. Acad. Sci. U.S.A. 73*, 3882–3886

Nau, H. & Riordan, J. (1975) Gas chromatography–mass spectrometry for probing the structure and mechanism of enzyme active sites. The role of Glu-270 in carboxypeptidase A. *Biochemistry 14*, 5285–5294

Tabushi, I., Shimokawa, K., Shimizu, N., Shirakata, H. & Fujita, K. (1976) Capped cyclodextrin. *J. Am. Chem. Soc. 98*, 7855–7856

Discussion

Cornforth: Could part of the catalytic activity of carboxypeptidase A be due to the substrate being bound in such a way that the amide group in question lies out of plane? I recall that when we were working on penicillin a young man called Woodward pointed out that the reactivity of such β-lactams could be explained by their reluctance to undergo the normal amide mesomerism.

Breslow: Lipscomb, who has done the X-ray structural work, has vacillated somewhat on the question of how much effect such torsion contributes. It does not seem to be a large factor.

Baldwin: Some people have suggested that those enzymes which so effectively destroy penicillin, the β-lactamases, are some sort of natural carboxypeptidase which happen to be good enzymes for that substrate.

Golding: Metal replacement studies (cf. Vallee & Williams 1968) support your analysis, Professor Breslow. Cobalt(II) carboxypeptidase shows peptidase activity, as one would expect.

Breslow: Cobalt(II) is kinetically labile, so substrate can replace the water and get into the coordination sphere, but cobalt(III) is inert to substitution.

Golding: Cadmium(II) carboxypeptidase has low peptidase activity but high esterase activity. This might be explicable on the basis of your arguments.

Breslow: We are saying, and at first this may seem a little uncomfortable, that enzymes can operate well by several mechanisms. It is not absolutely critical that carboxypeptidase use its normal mechanism; it can use a mechanism that is related but displaced by one molecule out. The normal mechanism is undoubtedly the better process; it is the only one in which the peptide comes close enough to the tyrosine for the proton to be transferred, but the tyrosine is not catalytic for all ester cleavages. Also, the zinc enzyme cleaves peptides and esters at about the same rate (if the substrates have similar structures). That means that the catalytic acceleration is about 1000-times less for esters than for peptides.

Prelog: One wonders why nature selected zinc, an ion that is not plentiful or widespread in the biosphere. Zinc is probably used often because it is a multipurpose ion. It can complex with different types of oxygen: carbonyl groups, carboxylate ions, hydroxy groups and water. In contrast, cobalt(III) ion complexes water much more strongly than carbonyl.

Breslow: Cobalt(III) cannot substitute for zinc. Cobalt(II) can, but unlike zinc it is subject to oxidation. One can make a case for carbonate dehydratase initially having a four-coordinate zinc hydroxide and building a five-coordinate zinc hydrogen carbonate. We have kept our proposal simple by assuming that the zinc maintains a four-coordinate state. Maybe life is not that simple!

Baldwin: If in the maleamic acid hydrolysis the phenolic OH group was masked and the reaction was done in $H_2^{18}O$, would the amide exchange with $H_2^{18}O$?

Breslow: Yes. In acidic conditions the problem is that the forward reaction of the tetrahedral intermediate is catalysed, otherwise there would be exchange of ^{18}O.

Baldwin: Do amides exchange ^{18}O at neutrality?

Breslow: Don't forget what kind of amide this is—it has a neighbouring carboxylate. In other words, it is an unusual and reactive amide. This catalytic carboxylate is the same catalytic group that we invoke for the ^{18}O exchange.

McCapra: Amides in acid solution are hydrolysed without ^{18}O exchange; the tetrahedral intermediate appears to decompose without reverting to the original amide.

Breslow: In base they do exchange ^{18}O. Our non-proton kinetics correspond to the basic conditions.

Ramage: In the maleamic acid hydrolysis how did the hydrolysis of the *N*-methylamide compare with that of the corresponding N–H compound?

Breslow: The *N*-proton plays no role in our scheme but we have not substituted it. Substituting a fused ring for the CO·NHR makes little difference.

Ramage: Hydrolysis of the normal polypeptide differs from that in which a *N*-methylamino acid is inserted.

Breslow: Yes; the carboxypeptidase does not hydrolyse a peptide with *N*-methylleucine as the terminal amino acid. *N*-Methylleucine does not catalyse ^{18}O exchange. This finding further corroborates the idea that the ^{18}O exchange proceeds by resynthesis of the starting material, which is a one-step process.

Kenner: Have you investigated the effect of introducing methyl groups at the maleic acid double bond? In the ordinary mechanism of hydrolysis, such substitution has a dramatic effect.

Breslow: We have not done that. We were not prepared for the change in mechanism. The nucleophilic mechanism is strongly promoted by bulky substituents on the maleamic acid double bond which squeeze these groups together. The general base mechanism might not be so promoted because it seems to have a more open geometry; we might see opposite substituent effects in this case.

Woodward: Do you see any way of improving the reverse reaction, so that the enzyme could synthesize peptides?

Breslow: If the methyl ester had been able to react with the amino acid, that would have been reaction in the exothermic direction, but the enzyme will not catalyse that reaction; from an organic chemical point of view it is remarkable that the amino acid cannot condense with a carboxylate ester but it can with a carboxylate anion (however, this specificity is probably due to the zinc). So, as it seems we cannot change starting material, we shall have to adopt another approach to drive the condensation, such as phase separation.

Kirby: The fact that the acidity of the phenol is not important, although the phenol is critical to the mechanism of the hydrolysis of the maleamic acid by acting as both an acid and a base, may mean that there should be an optimum activity for the phenol which depends on its pK. If it is either too strong or too weak it will be inefficient. Do you observe a pK optimum for a phenol when you vary its acidity?

Breslow: We have only tried a few cases.

Dewar: Have you used the amide instead of the methyl ester? The ester might not bind to the zinc whereas the amide would and the reaction would still be thermoneutral.

Breslow: D. Wernick has investigated many nucleophiles, including NH_3, to see if they could replace water. If the enzyme could catalyse the reaction of an amide with an amino acid one would see not the synthesis of the peptide but the hydrolysis of the substrate, since the peptide is a substrate for the enzyme. So the enzyme would synthesize the substrate, then cleave it. However, the amide was recovered unchanged.

McCapra: Does the cylodextrin catalysis involve a nucleophilic or general base mechanism?

Breslow: Unfortunately we know little about it. It catalyses the hydrolysis of amide with a base and an acid, but they are not typical bases or acids. It most resembles ribonuclease which also has two imidazole groups, one acting as a base, the other as an acid.

Reference

VALLEE, B. L. & WILLIAMS, R. J. (1968) Metalloenzymes: the entatic nature of their active sites. *Proc. Natl. Acad. Sci. U.S.A.* **59**, 498–505

General discussion II

SYNTHESIS OF ANTILEUKAEMIC LIGNANS

Raphael: Sir Robert was notable for the fact that he was the first or one of the first to think of many facets of chemistry. It is often forgotten that he first showed the synthetic potential of enamines by demonstrating that the enamine (1) derived from acetoacetate gave after methylation and aqueous work-up the

monomethylacetoacetate (2) exclusively (Robinson 1916). Enamine chemistry was neglected until Stork took it up in 1954. Since then, the use of enamines has mushroomed. One of the most useful properties of cyclic enamines such as (3) is their smooth reaction with dimethyl acetylenedicarboxylate in a blatant

non-Woodward–Hoffmann way to give a cyclobutene (4). Undoubtedly that process needs two steps and so the famous rules are not broken! On being heated, the bicyclic ring system rearranges to the substituted cyclooctadiene (5) which on aqueous hydrolysis with acid and monodecarboxylation yields the oxocyclooctenecarboxylate (6). It occurred to us that this well known process could be used effectively in the synthesis of the antileukaemic lignans, exemplified by the key compound steganone (7) (which can be transformed into the

remaining members of this class). Steganone contains a *trans*-fused γ-lactone, a methylenedioxy group and three methoxy groups and is biogenetically made up (as are the other lignans) of two C_6–C_3 units joined in this unusual way. Comparison of the substitution in the cycloctane ring in (7) with that in the enone (6) reveals the same –CO·C·C·CO·O– structure. For our synthesis of these lignans (Becker *et al.* 1977) we selected the substituted aminophenanthrene (8). We concluded that it was reasonable to suppose that 9-pyrrolidinophenanthrenes would behave in an enaminic way because the 9,10-double bond of phenanthrenes is ethylenic. On heating the aminophenanthrene (8) with dimethyl acetylenedicarboxylate, we obtained the ring-expanded product

(9) in over 90% yield. Acid hydrolysis and catalytic reduction converted this diester into the keto-acid (10) which gave the homogeneous pentacyclic lactone (11) (as a racemate) on treatment with base and HCHO at room temperature and oxidative work-up. Without a doubt, this was the *trans*-fused γ-lactone. Although we should have had the racemate of the compound corresponding to the natural product (7), the spectroscopic properties were completely different. The n.m.r. spectrum was considerably different and the i.r. spectrum was dramatically different, in particular the frequency of the benzylic

carbonyl group. This occurred at 1665 cm⁻¹ (as expected for a conjugated carbonyl group) in the naturally occurring compound but at 1710 cm⁻¹ in the synthetic racemate (i.e. the typical absorption for a non-conjugated carbonyl group). Then we found that when we heated the racemic compound (11) it rearranged quantitatively to a compound which was now identical spectroscopically with the naturally occurring compound.

The difference between the synthetic compound and the naturally occurring compound lies in the different directions of skew of the biphenyl group. The naturally occurring compound had the configuration (12), which was confirmed by the X-ray structure. The X-ray structure of a derivative of our synthetic compound showed beyond doubt that the skew of the biphenyl system was the other way (i.e. 13). The energy barrier between these two forms is sufficiently high to allow the isolation of the two compounds. A model of (12) shows that the carbonyl group lies in the plane of the benzenoid ring (this is confirmed by the X-ray structure), but a model of (13) shows that the carbonyl lies almost orthogonal to the plane of the benzenoid ring, so explaining the lack of conjugation between the carbonyl group and the ring. The quantitative nature of the rearrangement is probably due to the fact that full conjugation is attained in forming (12). The route to these potentially useful antileukaemic compounds—a 10-step synthesis from piperonal—proceeds with 23% overall yield.

There is a Robinsonian precedent to this stereochemical arrangement (the first example observed in natural product chemistry) in dihydrothebaine (14) which has an o,o-junction with a nine-membered ring. Sir Robert showed that this was optically active owing to the presence of the biphenyl system.

Sondheimer: You initially made the analogue of steganone with completely unsubstituted benzene rings. Did you find the same sort of isomerism?

Raphael: No. Before we embarked on this synthesis we studied the process starting from the simple parent 9-pyrrolidinophenanthrene but found no indication of this phenomenon, probably because the energy barrier to the rearrangement analogous to (13) → (12) is considerably higher in the substituted case on account of the *o,o*-interaction between the buttressed *o*-methoxy group and the *o*-hydrogen (see 15). I should like to return to the saturated keto-acid (10) obtained by catalytic hydrogenation. This process had given only one homogeneous compound but when we discovered the above thermal rearrangement we tried to rearrange (10) thermally. This treatment gave us a 50/50 mixture of the unchanged compound and a new isomeric keto-acid, formaldehyde–base treatment of which yielded the naturally occurring steganone. Thus we had changed the sense of the skew by isomerization of (10).

ORGANOMETALLIC SYNTHESIS

Birch: Sir Robert was often more interested in new synthetic methods to solve structural problems than in carrying out the final syntheses. I am sure he would have enjoyed an area to which I want to draw the attention of organic chemists involving some elements in the Periodic Table which they do not use but which can assist their synthetic endeavours. I believe that organometallic chemistry will in the future provide a major series of synthetic processes and I shall illustrate this with some examples of reactivity related to tricarbonyl iron–diene complexes (see, e.g., Birch & Williamson 1973; Birch *et al.* 1973a,b).

The $Fe(CO)_3$ group has several important effects. It totally distinguishes one side of the molecule from the other, particularly in cyclic compounds, and thus enables us to do totally stereospecific syntheses—some reactions proceeding

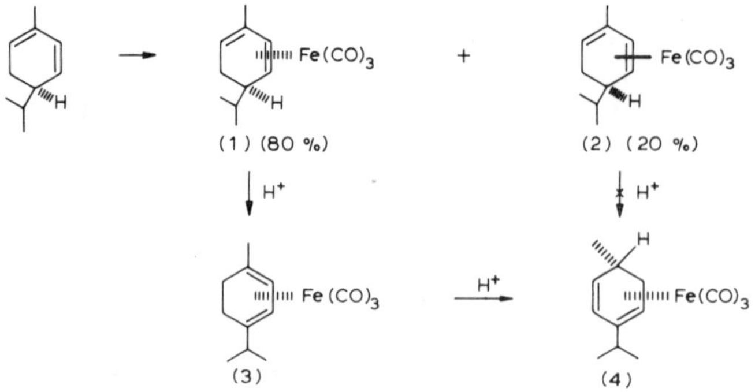

SCHEME 1

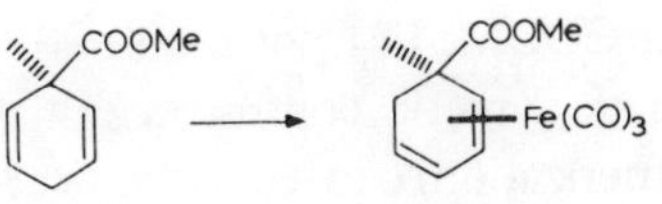

ORGANOMETALLIC SYNTHESIS

SCHEME 2

on the same side, others on the opposite side. It also stabilizes unstable organic products. Using the addendum we can synthesize optically active compounds in new ways. It provides specific methods for isomerizing double-bond systems, and also enables molecules to be sensitized effectively to nucleophilic attack. The $Fe(CO)_3$ group can be finally removed.

Scheme 1 illustrates stereospecific attachment to the rear of a substituent as the result of steric hindrance, to one natural enantiomer of α-phellandrene. Scheme 2 depicts intermediate complexing with a substituent which directs to the same side as the substituent. For stereospecific migration with acid to occur the H which moves must be on the same side as the $Fe(CO)_3$ group. The menthadiene complex (1) is isomerized to (3) and (4) after protonation on Fe, but the enantiomeric complex (2) merely racemizes. The products (3) and (4) are optically active.

On formation of the $Fe(CO)_3$ complex, the least hindered side of a diene then becomes the most hindered side, thereby allowing unusual reactions, for example reduction of the ergosterol derivative (5) to the 3α-OH. Direct reduction of the uncomplexed ketone gives the 3β-OH.

Formation of the complex allows specific isomerization (Scheme 3) in ways not possible by other known procedures which give transoid dienes or disproportionation products.

In the acid-catalysed isomerizations of this type saturated substituents are

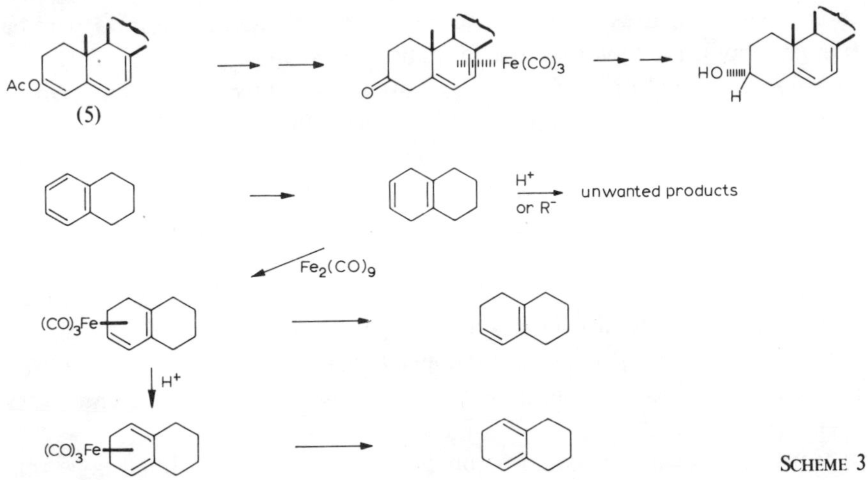

SCHEME 3

finally found attached to 2-positions and ones like COOMe to 1-positions of the complexed diene, leading to predictable products. Kinetic control in initial complexing or later thermodynamic control (isomerization) can lead specifically to different products as shown in Scheme 3.

Hydride abstraction from neutral complexes leads to stable carbonium salts usually in the form of fluoroborates. Such carbonium salts undergo nucleophilic attack from the opposite side of the ring to the $Fe(CO)_3$ (e.g. Scheme 4).

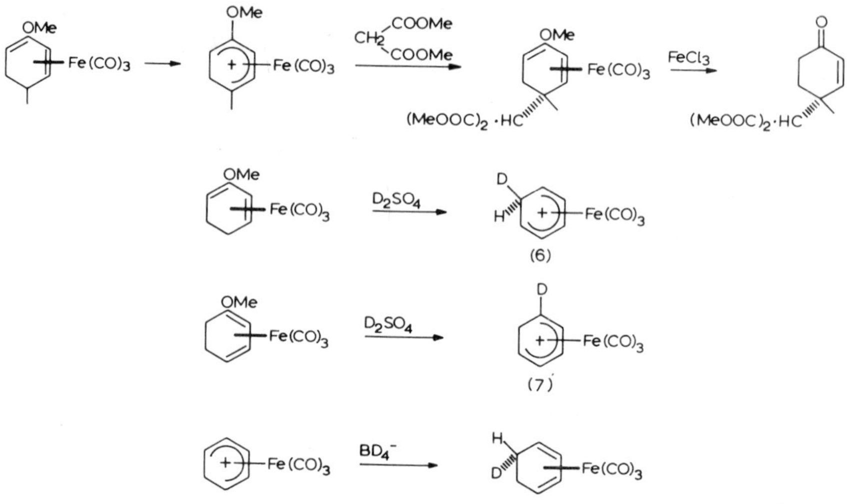

SCHEME 4

Such salts can be generated in other ways by acid treatment of neutral methoxydiene complexes. Specific incorporations of deuterium from D+ can be brought about. Examples are the formation of (6) and (7).

Among the nucleophiles that can be used BD_4^- is useful; its deuterium is incorporated stereospecifically (e.g. formation of [8]).

As well as addition (or loss) of a proton by attachment in an intermediate to the Fe atom, probably as $-FeH(CO)_3^+$, some other reagents attack the β-side of the ring probably through attachment to Fe (e.g. acetylation of the diene complex [9]).

An example of the stabilization of unstable molecules is the complex which is made indirectly with cyclohexadienone, the ketonic form of phenol. Although the carbonyl group is rather unreactive, this complex (10) can be used as a phenylating agent either in acid or by transformation into the enol-ether (11).

Many other examples could be quoted, and other metals, but I hope that these will suffice to alert the organic chemist to the fact that he must look into the organometallic literature. Indeed, he has also a specific contribution to make since, unlike some inorganic chemists, he is not frightened by substituents. In turn he must learn not to be frightened by metal atoms.

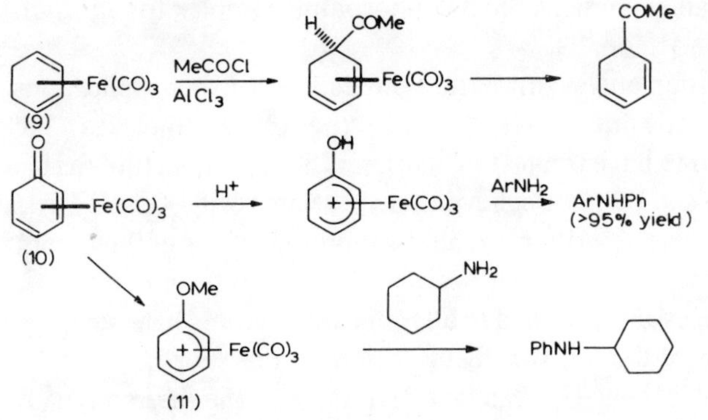

Raphael: One snag is that removal of the Fe(CO)₃ group is sometimes difficult. Do you have a method that will guarantee good yields in every case?

Birch: The best reagent seems to be trimethylamine oxide, Me₃NO. We have also used iron(III) chloride (Birch et al. 1973) and copper(II) chloride. The choice rests on trial and error for each compound.

Brown: A short column of activated MnO_2 works well for cycloheptatrienes.

Birch: When we used that in one case (Scheme 5) the MnO_2 apparently removed electrons from the iron and oxidized our product by removing H from the same side leading to ring closure on the other side.

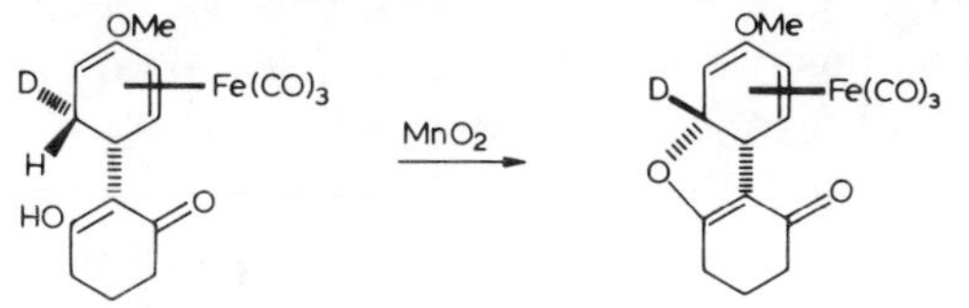

SCHEME 5

Sondheimer: Fe(CO)₃ is extremely useful for stabilizing certain exceedingly unstable dienes such as cyclobutadiene.

Golding: Professor Birch, you could prove the stereochemistry of the protonation (deuteriation) of (1) to (2) by converting monodeuteriated (2) into [²H]-succinic acid and comparing its o.r.d. spectrum with the published spectra (Cornforth et al. 1962) for (2R)- and (2S)-[²H]succinic acid.

Brown: In principle, one can use these isomerizations starting with a chiral unresolved diene and an asymmetric metal–phosphine complex for an asymmetric synthesis.

Birch: I agree; resolution and asymmetric synthesis are further applications.

Sondheimer: A semantic question is how one writes these complexes? Victor *et al.* (1970) claimed to have trapped two different Kekulé structures as bistricarbonyliron complexes. I do not believe that these are really Kekulé structures, because the formulae as written are purely representations of these complexes.

Birch: The iron is necessarily bonded to all four carbon atoms in a diene system; that explains some of the unusual features of reactions. For example, the thermodynamically stable diene (12) can be converted into the thermodynamically unstable isomer (13) since the latter can be stabilized in the form of the

intermediate complex. In this way, one can force a reaction up the energy hill.

A major deficiency in this area at present is lack of a full understanding of what this bonding means in terms of reactivities and interactions with attached groups. ^{13}C n.m.r. evidence in unpublished work by A. Pearson (Canberra) seems to indicate that the σ–π bonding, as in (14) for instance, is not an unreasonable representation, although this had previously been rejected on the basis of ^{1}H n.m.r. spectra. How substituents such as MeO, Me, MeOOC and so on exert their effects on reactivities is difficult to say. For example, it is not clear why treatment with acid of a cyclohexadiene–Fe(CO)$_3$ complex which has a methyl substituent attached anywhere to the ring (15) always gives the

2-methyl product (16). Clearly, that is the thermodynamically stable isomer; but why? In contrast, with a methoxycarbonyl substituent (17), the 1-substituted product (18) is obtained. Rationalization of the effect of MeO on reactivities of the complexed diene system is not easy. It does not seem to react by any kind of mesomeric effect; probably its inductive effect is dominant.

Dewar: With regard to Kekulé structures, one can fix them: for example Davis & Pettit (1970) isolated the complex (19) which the X-ray crystal structure showed has bonds absolutely localized in the benzene ring.

Rees: That exemplifies the case when one of the double bonds of the diene system is part of an aromatic ring. Are any examples known in which one of the double bonds is part of an allenic system?

Birch: With $Fe_2(CO)_9$, allene itself gave the $Fe(CO)_4$ complex (20). Such $Fe(CO)_4$ complexes are produced with several olefins and involve acceptance of only two π-electrons. They are much more labile than the $Fe(CO)_3$ complexes and their reactions have been little studied. They are probably intermediates to diene–$Fe(CO)_3$ complexes.

Brown: Several examples of hydride migration are known. Have alkyl or acetyl migrations (of, say, Friedel–Crafts products) been observed? Since the C–H bond can be activated, a C–C bond might be.

Birch: I know of no such migrations. Products of nucleophilic reaction, however, such as the malonate (21), resist further attempts to remove hydride; the malonic ester group instead is lost. This is a simple breakage of a C–C bond. Even a $CH_2 \cdot CO \cdot CH_3$ substituent in the 4-position is removed by H_2SO_4 to regenerate the cation.

Cornforth: Do these complexes ever undergo Cope-type rearrangements?

Birch: I see no reason why they should not.

Brown: Metal complexation can accelerate (Heimback & Brenner 1967), retard (Moriarty *et al.* 1972) or redirect (Aris *et al.* 1975) a Cope rearrangement.

Baker: We have made the bis-allylic dimer (22) from isoprene and Ni(0). In our studies we have been interested in electrophilic and nucleophilic additions to π-allyl groups. The addition depends on the metal and the ligand. We have demonstrated regioselective addition of both electrophiles and nucleophiles (Baker *et al.* 1975). The bis-allyl intermediate may exist as the σ–π form

(23). Nucleophiles (N) preferentially attack the σ-form and electrophiles (E) attack the π-form. We have observed parallel reactions with the bis-π-allylnickel complex (24), obtained from allene, which also appears to exist in the σ–π form, and again we see regioselective attack.

Although we have not yet used the reactions of (24) to build terpenes in good yield, this dual reactivity of bis-π-allylnickel intermediates has been useful in other ways. For example, the bis-π-allylmetal intermediate (25) can be obtained from butadiene with either nickel or palladium catalysts. This reacts at the elec-

trophilic carbon atom of phenylhydrazones and after a hydride shift yields azo compounds (26) and (27) (Baker *et al.* 1976). Initial reaction can also take place at the 3-position of the allyl derivative. In some conditions attack can also take place at the nucleophilic centre of the hydrazone and a C_8 alkylation occurs.

With oximes, reaction occurs at the 3-position of the bis-π-allylmetal complex to yield the divinyloxazacycloheptane (28) which involves both electrophilic and nucleophilic attack (R. Baker & M. S. Nobbs, unpublished work).

(25)   R₂C=NOH →   →   (28)

This selectivity of reaction with bis-π-allylmetal intermediates could be useful for the synthesis of alicyclic and heterocyclic derivatives.

ANTI-AROMATIC COMPOUNDS

Kenner: An area we have not yet considered is the use of anti-aromatic compounds as intermediates in synthesis.

Rees: There is a modest number of papers on the use of anti-aromatic compounds as intermediates in carbocyclic chemistry, based largely on cyclobutadiene as a ligand in transition metal complexes. Much less has been reported with anti-aromatic heterocyclic compounds. The azacyclobutadiene (azete) system which we have made fused to a benzene ring displays the expected high reactivity in nucleophilic additions and cycloadditions (Adger *et al.* 1975). However, attempts to isolate heterocyclic anti-aromatic systems like this one as ligands on transition metals appear to have been wholly unsuccessful.

References

ADGER, B. M., REES, C. W. & STORR, R. C. (1975) Benzazetes (1-azabenzocyclobutanes). *J. Chem. Soc. Perkin Trans. I*, 45–52

ARIS, V., BROWN, J. M., CONNEELY, J. A., GOULDING, B. T. & WILLIAMSON, D. H. (1975) Synthesis and thermolysis of rhodium and iridium complexes of *endo*-6-vinylbicyclo[3,1,0]hex-2-ene. A metal promoted vinylcyclopropane to cyclopentene rearrangement. *J. Chem. Soc. Perkin Trans. II*, 4–9

BAKER, R., COOK, A. H. & CRIMMIN, M. J. (1975) Regiospecificity in the reactions of bis-π-allyl-nickel intermediates with nucleophilic and electrophilic reagents. *J. Chem. Soc. Chem. Commun.*, 727

BAKER, R., NOBBS, M. S. & ROBINSON, D. T. (1976) Reactions of phenylhydrazones with bis-π-allyl-nickel and -palladium complexes. *J. Chem. Soc. Chem. Commun.*, 723

BECK, D., HUGHES, L. R. & RAPHAEL, R. A. (1977) Total synthesis of the antileukaemic lignan steganacin. *J. Chem. Soc. Perkin Trans. I*, in press

BIRCH, A. J. & WILLIAMSON, D. H. (1973) Organometallic complexes in synthesis. Part V. Some tricarbonyliron derivatives of cyclohexadienecarboxylic acids. *J. Chem. Soc. Perkin Trans. I*, 1892–1899

BIRCH, A. J., CHAMBERLAIN, K. B., HAAS, M. A. & THOMPSON, D. J. (1973a) Organometallic complexes in synthesis. Part IV. Abstraction of hydride from some tricarbonylcyclohexa-1,3-dieneiron complexes and reactions of the complexed cations with some nucleophiles. *J. Chem. Soc. Perkin Trans. I*, 1882–1891

BIRCH, A. J., CHAMBERLAIN, K. B., & THOMPSON, D. J. (1973b) Organometallic complexes in synthesis. Part VI. Some oxidative cyclizations of tricarbonylcyclohexadieneiron complexes. *J. Chem. Soc. Perkin Trans. 1*, 1900–1902

CORNFORTH, J. W., RYBACK, G., POPJAK, G., DONNINGER, C. & SCHROEPFER, G. (1962) Stereochemistry of enzymatic hydrogen transfer to pyridine nucleotides. *Biochem. Biophys. Res. Commun. 9*, 371–375

DAVIS, R. E. & PETTIT, R. (1970) Bond localization in aromatic-iron carbonyl complexes. *J. Am. Chem. Soc. 92*, 716–720

HEIMBACK, P. & BRENNER, W. (1967). *Angew. Chem. Int. Edn. Engl. 6*, 800

MORIARTY, R. M., YEH, C.-L., YEH, E.-L. & RAMEY, K. C. (1972) The tungsten pentacarbonyl complex of semibullvane. Barrier height for the degenerate Cope rearrangement. *J. Am. Chem. Soc. 94*, 9229–9230

ROBINSON, R. (1916) The C-alkylation of certain derivatives of β-aminocrotonic acid. *J. Chem. Soc. 109*, 1038–1046

VICTOR, R., BEN-SHOSHAN, R. & SAREL, S. (1970) Trapping of Kekulé structures *via* co-ordination to iron. Positional isomerism in bis-tricarbonyliron complexes of 3,α-dimethylstyrene. *J. Chem. Soc. Chem. Commun.*, 1680–1681

Summing up

LORD TODD
Christ's College, Cambridge

In my introductory remarks I referred to the wide range of topics we were to discuss and indeed it has been wide; but the Robinsonian thread has run through them all. Professor Birch's talk reflected and updated Sir Robert's speculations on biosynthesis and demonstrated the use of biogenesis as an aid to structural studies. Professor Battersby described elegant work on biosynthesis and this tied in naturally with much discussion on enzymes and nature's economy. Porphyrin biosynthesis is an excellent starting-point for argument — why should nature particularly want a type-III porphyrin for so many purposes? Interesting in this connection was the suggested analogy between gas-phase reactions and those of enzymes. The mechanism of enzyme action was indeed the subject of a good deal of rather inconclusive discussion.

On synthetic methods, in addition to polyketide reactions we had an account of work on the development of the Robinson ring annelation methods and the Robinsonian connection was further stressed through Professor Ramage's reference to the isoprene rule in biogenesis. I am sure that Professor Barton's paper on specific substitution in phenolic compounds would have fascinated Sir Robert and his use of selenium emphasizes Professor Birch's point (p. 194) about the way in which elements which would have been regarded as most unusual 20 years ago are being increasingly applied in organic synthesis.

If I understand Professor Dewar aright, he feels that his calculations are now reaching an accuracy sufficient to allow useful predictions about some reactions. This is a considerable advance although I confess to a certain scepticism. I fear that we may only be reaching the kind of accuracy required to explain reactions we already know but not enough to give any precise information about those which are novel and interesting.

Substantial changes have shaken organic chemistry since Sir Robert was active. He used to dislike the new physical methods and frequently claimed that they would stop the development of the subject. He was certainly wrong about that, but it is true that the rise of instrumentation in the past 20–25 years has radically changed the way in which most people approach the subject. Before that time the determination of structure was a long and arduous task but it added continuously to our knowledge of reactions; so too did synthesis, which was pursued very much as an adjunct to structural work in order to verify its conclusions. Today the development of physical methods has enormously reduced the amount of degradative work and has caused synthesis to diminish in importance as a verification of structure, although of course it remains vigorous in what might be called its technological aspect — that is, its use in preparing particular compounds for study or use. It has long seemed to me that development along these lines was bound to occur and that, as it did, people's minds would move from a consideration of structure *per se* to structure in relation to function especially in biological systems. Our discussions appear to me to have underlined this view.

Running through much of our discussion during this symposium has been a striving to understand heterogeneous catalysis of the type exhibited by natural enzymes and we have had a number of contributions indicating progress in this field. As I have already suggested (p. 103) we may soon know enough about enzymes to use their methods without having to use the natural products themselves. Thus, Dr Brown's work on micelles shows how peptides may be combined with different materials some of which may not be just solid supports but actually part of a reacting system. Again, work on synthesis seems now to be aimed in certain areas at understanding mechanisms which are of biological significance; Professor Baldwin's rules on ring closure have a bearing on this point.

The study of structure in relation to function marks perhaps a point of departure or perhaps just a development from the work of Sir Robert. He did not, as far as I am aware, devote much thought to the function of natural products; that was something for his successors, and this symposium seems to me to have significantly emphasized this.

Index of contributors

Entries in **bold** *type indicate papers; others refer to contributions to discussions*

Baker, R.   79, 93, 96, 97, 171, 200
Baldwin, J.   22, 65, **85**, 92, 93, 94, 95, 96, 97, 98, 105, 124, 126, 127, 146, 171, 172, 186, 187
Barton, Sir Derek   22, 44, 45, 46, 47, 48, 50, **53**, 60, 61, 62, 63, 64, 65, 80, 81, 97, 101, 102, 104, 105, 124, 125, 145, 146, 147
Battersby, A. R.   20, **25**, 42, 43, 44, 45, 46, 47, 49, 105, 145, 172
Birch, A. J.   **5**, 20, 21, 47, 81, 82, 144, 147, 194, 197, 198, 199
Breslow, R.   21, 43, 46, 48, 63, 94, 98, 127, 129, 171, **175**, 186, 187, 188
Brown, J. M.   24, 63, 93, 94, 126, **149**, 171, 172, 173, 197, 199, 200

Chain, Sir Ernst   102, 103, 105
Cornforth, Sir John   21, 43, 45, 65, 99, 124, 186, 199

Dewar, M. J. S.   22, 23, 24, 43, 45, 48, 60, 61, **107**, 122, 123, 124, 125, 126, 127, 129, 146, 147, 172, 173, 188, 199

Eschenmoser, A.   23, 45, 103
Evans, G. E.   **131**

Garson, M. J.   **131**
Golding, B. T.   22, 44, 46, 65, 94, 95, 96, 103, 147, 186, 197

Griffin, D. A.   **131**

Jones, Sir Ewart   81

Kenner, G. W.   **1**, 42, 101, 102, 104, 105, 145, 147, 173, 188, 201
Kirby, G. W.   42, 47, 50, 188

Leeper, F. J.   **131**
Ley, S. V.   **53**

McCapra, F.   21, 42, 122, 123, 124, 171, 187, 188

Prelog, V.   63, 82, 97, 102, 103, 187

Ramage, R.   24, 44, 46, 47, 49, **67**, 80, 81, 82, 97, 104, 147, 187
Raphael, R. A.   42, 60, 64, 191, 194, 197
Rees, C. W.   61, 62, 63, 125, 126, 199, 201

Sondheimer, F.   105, 194, 197, 198
Staunton, J.   43, 45, 47, 48, 49, 50, **131**, 145, 146, 147

Todd, Lord   **3**, 64, 82, 101, 102, 103, **203**

Woodward, R. B.   22, 23, 43, 45, 46, 50, 63, 92, 96, 97, 101, 102, 103, 104, 122, 125, 126, 188

Indexes compiled by David W. FitzSimons

Subject Index

acetalization
  91, 93, 96
acetate
  acetyl-CoA 12, 14, 145, 147
  head-to-tail linkage 12
  incorporation of 8, 13
  malonyl-CoA 14
  polyketide hypothesis 12, 14, 147
  porphyrin side-chains 43
acorenol
  72, 73, 74
acoric acid
  5, 6
activation energies
  110, 112, 114, 115, 117-119, 123
agarospirol
  71, 74, 75
alkaloids
  benzophenanthridine 27, 28
  benzylisoquinoline 26
  biosynthesis 26–31
  function 45
  1-phenethylisoquinoline 30, 31, 46
  strychnine 82, 83
alkyl halides
  23, 114
π-allyl-nickel complexes
  200, 201
amides
  bond cleavage 176-185,
  187, 188
  ^{18}O exchange 187
  synthesis 101, 104, 105, 181, 182, 187
  and zinc 188
δ-aminolaevulinate
  94, 95
α-amyrin
  21
anthracenes
  132, 136–140
anthraquinones
  13, 144
anti-aromatic compounds
  201
arene oxides
  49, 50

benzyne
  113, 126
beryllium borohydride
  110, 112
bilane
  37, 38, 39, 42, 43
biomimetic synthesis of phenols
  131–147
biosynthesis
  biomimetic syntheses 131–147
  economy of nature 27, 31, 44, 45, 46, 49
  morphine 26–29, 44, 46, 50
  of polyketides 147
  polyketone cyclization 139, 144–147
  porphyrins 31–40, 42, 43
  propionate 14, 46
  Sir Robert Robinson's ideas 2, 25, 26
  steroids 22, 68
  structure determination 5–24
biradicals
  benzyne 113
  bond length 126
  butanal decomposition 127, 128
  Cope rearrangement 119, 120, 124, 125, 126, 127
  coupling 47, 50
  cyclopropane ring formation 127
  Diels–Alder reaction 115, 116
  dioxetan dissociation 122, 123
  geometry 125, 126
  MINDO/3 113, 126–128
  in phenol oxidation 47–50
  reactivity 124
  and ring closure 93, 124, 125
  ring opening 118
  sigmatropic rearrangements 126
bond angles
  109, 110

bond lengths
  in biradicals 126
  in Diels–Alder reaction 116
  iminoquinones 61
  MINDO/3 calculations 109, 110, 116
  selenoimines 60, 61, 62
  thioimines 60, 61, 62
brevianamide-A
  10, 11

carbenes
  21, 22, 109
carbonium ions
  in biosynthesis 6–9, 21, 24
  and carbenes 21, 22
  cyclopropane ring formation 21, 22
  after hydride extraction 21, 196
  and spirosesquiterpenes 75
  tricarbonyliron-diene complexes 196
carboxypeptidase A
  amide cleavage 175–180, 184
  ester hydrolysis 183, 184
  β-lactamases 186
  and maleamic acid derivatives 177–180, 187
  mechanism of action 176–185, 187
  methanol 182, 183
  metal replacement studies 186, 187
  peptide synthesis 181, 182, 187, 188
  phenolic OH 177–181, 183, 187, 188
  structure 175, 176
  role of water 178, 180, 182, 183, 184
  zinc 176, 177, 184, 186, 188
catalysis
  acyl transfer in micelles 161–164
  alkylation of naphthoxide 167–168
  asymmetric 157–160

  bromination of anisole 172, 173
  cationic micelles 164, 165
  enantiomers 159, 160, 171
  ester hydrolysis 151, 152
  micellar 152–173
  and monolayers 151
  ^{18}O exchange 182
  of peptide hydrolysis 175
  $Ru^{II}(bipy)_3$ complexes 151
  selective homogeneous and heterogeneous 149–173
  solid-state 171
α-cedrene
  72, 73, 74
cerulenin
  18
chymotrypsin
  23, 182
colchicine
  biosynthesis 29–31
Cope rearrangement
  119, 120, 124, 125, 126, 127, 199, 200
corticotropin
  synthesis 102
cycloartenol
  22
cyclobutadiene
  113, 197, 201
cyclodextrin
  184, 185, 188
cyclopropanes
  21, 22, 109, 127
α-cyperone
  70, 88

1,2-dehydropyrrolizidine alkaloids
  courtship pheromones 45, 46
Diels–Alder reaction
  81, 115–118
diphenylseleninic anhydride
  chemistry 54, 65
  hexamethyldisilazane 60, 63
  phenol oxidant 54–60, 61, 62

echinulin
  10, 21

economy of nature
  27, 31, 44, 45, 46, 49, 146
enamines
  191–194
enzymes
  action 22, 102, 103
  artificial 102, 103, 184
  carbonium ion formation 22
  carboxypeptidase A 175–188
  and catalysis 149, 150, 182
  chymotrypsin 23, 182
  cosynthetase 32, 34, 39, 42, 43
  deaminase 32, 34, 39, 42, 43
  epimerases 46
  folding of protein 182
  β-lactamases 186
  mechanism of action 22, 23, 102, 103, 150, 176, 178, 180, 182, 186, 188
  and micelles 153
  models 102, 103, 105, 153, 175–188
  ^{13}C n.m.r. 155, 156
  oxygen+benzene reaction 124
  paramagnetic probes 155, 156
  peptide synthesis 181, 187
  and phenol oxidation 47
  polyketide cyclization 139, 144, 145, 146
  in porphyrin biosynthesis 32, 34, 39, 42, 43
  and protective groups 45
  selectivity 150
  solid-phase synthesis 101, 102
  solvent 22, 23, 24, 42, 178, 180, 182, 183
  stereochemistry 46, 103
  steroid biosynthesis 22
  terpene biosynthesis 22
  trypsin 182

flavonoids
  47
free energy
  biradicals 125–128

cyclopropane ring formation 127
in micelles 153
MINDO/3 calculation 119, 123

gas-phase reactions
22, 23, 24, 114
genetic engineering
105

hexadiene
119, 120, 124–126
hexamethyldisilazane
60, 63
histidine
in chymotrypsin 23
histidine esters
acyl transfer in micelles 161–164
asymmetric catalysis 157–160, 171
hydroxylation
aryl rings 49
diphenylseleninic anhydride 54–60, 62, 64
mechanism 54–60, 65
tetracycline ring A model phenols 54, 55, 65

insulin
synthesis 102, 105
iron carbonyl
(see also tricarbonyliron)
93, 96
isoprene
Ni(0) dimers 200
rule 7, 67, 71

lignans, antileukaemic
synthesis of 191–194

maleamic acid derivatives
177–180
malonyl-CoA
14, 46, 144
5-membered heterocycles (O, N, S, Se)
bonding 60–62
formation 88–92, 93, 94, 96, 97, 109, 201

stereoelectronics 93, 94, 97
structure 60, 61, 62
C-**methylation**
20
O-**methylation**
in flavonoids 47
in morphine biosynthesis 44, 46, 47
and phenol oxidation 44, 47
protective groups 4, 45, 46
of polyphenols 139, 146
methymycin
14
micelles
acyl transfer 161–164
catalysis 153, 161, 162, 164, 165
cationic 164, 165, 166
formation 152
heterogeneous surfaces 166–168
structure 153, 155–157, 171
Michael reaction
64, 69, 86, 88, 89, 90, 97
MINDO/3
bond properties 110
configuration interaction 129
Cope rearrangement 119, 120, 126
development 108
Diels–Alder reaction 115–118
dioxetan dissociation 122, 123
for peroxides 124
reaction mechanisms 112–121
selenoimine structure 61
thermodynamic quantities 109–112, 123
transition states 114, 115, 116
vibration frequencies 110, 112
morphine
biosynthesis 26–29, 44, 46, 47, 50
and enkephalin 102
O-methylation 44, 46, 47

naphthacenes
139, 140
naphthalenes
136–140, 142, 143
naphthoxide ion
167, 168
neocembrene
6
nickel
π-allyl complexes 200, 201
nitrenes
62
nootkatone
71, 81
nystatin
biosynthesis 14, 18

oligonucleotide synthesis
101
organometallic synthesis
194–201
oxene
49
oxidation
CrO_3 21
of *gem*-dimethyl groups 20
Kuhn-Roth 8, 9, 15, 20
lead tetraacetate 14, 62
of phenols 27, 44, 47, 50
of pleuromutilin 8, 9
tetracycline ring A model phenols 53–60
S-**oxides**
62, 124
oxygen
alkylation 44–47, 139, 146, 167, 171
compounds, MINDO/3 calculations for 109
exchange reactions 180–182, 187
in flavonoids 47
hydroxylation of phenols 54–58
in phenol oxidation 47–50
reaction with benzene 124
reaction with olefins 113, 123, 124
singlet 113, 114, 123
in terpenes 45
zinc complexes 186

penicillin
 85, 86, 87, 93, 186
peptides
 hormones 102, 105
 hydrolysis by carboxypeptidase A 175–184
 hydrolysis by cyclodextrins 184, 185
 synthesis 101–106, 181, 182, 187
peroxirans
 113, 114, 123, 124
phenanthrenes 139
phenol oxidation
 27, 47, 53–65
 arene oxides 49
 biradicals 47, 48
 of cathylates 55, 56, 65
 diphenylseleninic anhydride 54–60
 hexamethyldisilazane 60
 mechanism 47, 48, 50, 57, 58, 62, 65
 methylation and 44
 phenolate 56
 phenoxonium system 47–50
phenols
 and amide cleavage 177–181, 183, 187, 188
 amination 60, 64
 biomimetic synthesis 131–147
 coupling 27, 48, 49, 50
 ketonic form 196
 methylation 44, 47, 64, 139
 oxidant for 53–65
 phenolate anion 48, 49, 55, 56, 57, 61
 phenylselenation 57, 60, 61
 from polyketone cyclization 135, 136, 137, 139, 140, 141, 142
 tricarbonyliron complex 196
phenylselenium nitride (PhSeN)
 62, 63
phomazarin
 11–14

pleuromutilin
 7–9, 20
polyketide hypothesis
 acetate 12, 14
 anthracenes 132, 136–140
 naphthalenes 136–140, 142, 143
 nystatin biosynthesis 14–18
 phenanthrenes 139
 phomazarin biosynthesis 11–14
 polyketone cyclization 135–147
 polyketone formation 133–135
 propionate 14
 pyrones 134, 135, 136, 140, 141
 Sir Robert and 132
polyketones
 and C-alkylation 140
 aromatic ring formation 132, 136–140, 142–144
 and biphenyl 142, 143
 curvularin 145
 cyclization 135–147
 cis-enol ethers 140, 141, 145, 146
 folding 138, 140, 141, 147
 formation 133–135, 140, 141
 malonyl-CoA 144
 O-methylation 139, 146
 phloroglucinol 135
 polyketide hypothesis 132, 133
 resorcylic acid 135, 140
porphobilinogen
 bilane formation 37
 biosynthesis 94, 95
 combination 31, 32
 porphyrin biosynthesis 33, 34, 36, 43
 rearrangement 34–39
porphyrins
 biosynthesis 31–40, 42, 43
 separation 37
propionate
 14–18, 20, 46
protecting groups
 4, 45, 46, 47, 73, 74
protein synthesis
 101, 102, 103, 104

pyrones
 polyketone formation 133, 134, 140, 141, 142
 pyranopyrones 134, 135
 rearrangement 135, 136, 142, 143

quantum organic chemistry
 107–109, 118

radicals
 (see also biradicals)
 benzyne 113
 Cope rearrangement 119, 120, 124, 125, 126, 127
 interception 124
 and micelle structure 155
 and MINDO/3 calculation 109, 115, 116
 in phenol coupling 48, 49
rate constants
 116, 117, 182
reaction kinetics
 amide hydrolysis 178, 179, 182
 asymmetric catalysis 157–160
 bromination of anisole 172, 173
 carboxypeptidase A 178, 179
 Diels–Alder reaction 117
 in micelles 153, 158
 MINDO/3 116
 ^{18}O exchange 182
 surface-layer catalysis 151
reaction mechanisms
 acyl transfer in micelles 161–164
 asymmetric catalysis 157–160
 bromination of anisole 172, 173
 butanal decomposition 127, 128
 carboxypeptidase A models 177–184
 cationic micelles 164, 165
 Cope rearrangement 119, 120, 124, 125, 126
 Diels–Alder reaction 115–118

dioxetan dissociation 122, 123
energy surfaces 123, 127, 128
on enzymes 149, 150
ester hydrolysis 151, 152, 183, 184
maleamic acid hydrolysis 177, 180
micelle-like behaviour 172, 173
MINDO/3 calculation 112–121, 123, 124, 127, 128
pentadiene rearrangement 127
peptide hydrolysis 176, 177
peptide synthesis 181, 182, 187
peroxiran formation 113, 123, 124
phenol coupling 27, 48, 49, 50
polyketone cyclization 135–147
regioselective attack on allyl-nickel complexes 200
sigmatropic rearrangements of ylides 126
ring annelation (Robinson) 68–75, 79, 81
ring closure
85–92, 96, 97, 98
carbanion additions 89, 96
enamines 88, 95
energy of 127
hydroxyenones 88, 89, 90, 91
mechanism 127
5-membered rings 88–92, 93, 94
in penicillin synthesis 86, 87
porphobilinogen biosynthesis 94, 95
radicals 127
rules for 85–92, 96, 97, 98
S_N2 process 96, 97
squalene epoxide 82, 147
stereoelectronics 87, 91, 93–96, 99
transition state 87, 98, 99, 127

ring contraction
72
ring expansion
29, 191–194
ring opening
118
Sir Robert Robinson
alkaloids 82
anthracenes 137
biomimetic synthesis 131, 137
and biosynthesis 2
biphenyl isomerism 193
colchicine 29
and colour 3
α-cyperone 70, 81
emodin 137
enamine chemistry 191
endocyclic transfers 85
eremophilane 67, 71
δ-hexenolactone 81, 82
laboratory equipment 4
morphine biosynthesis 26, 45
polyketide biosynthesis 132, 137
porphyrins 31, 33
protecting groups 45
ring annelation 68–75, 79, 81, 82
sesquiterpene biosynthesis 67
and solid-phase synthesis 101, 102
steroid synthesis 68, 69
strychnine structure 82, 83
synthetic method 3, 4, 194
theoretic organic chemistry 107
tricarbonyliron-diene complex 194
tropinone synthesis 131, 132

selenoimines
60, 61, 62, 63, 64
sesquiterpenoids
(see also specific compounds)
dihydroagarofuran 71, 79
eremophilone 67, 71, 81
Robinson ring annelation 68–75, 79, 81
spiro 72–75

spirovetivones 72
stereochemistry 71–75
synthesis 67–76, 79, 80, 81
zizanoic acid 71, 72
solid-phase synthesis
101–104
solvent
carboxypeptidase A 178, 180, 182
in enzyme reactions 22, 24, 42, 178, 180
in gas-phase reactions 23, 24
and peptide synthesis 182
squalene
68, 82, 147, 149
steganone
192, 194
stereochemistry
alignment of reactants 87, 93, 96, 97
asymmetric micellar catalysis 157–160
and biogenetic isoprene rule 71
chirality 103, 158, 168, 171, 197, 198
Cope rearrangement 119, 120
Diels–Alder reactions 81, 114, 116, 117, 118
enantiomeric recognition 159, 160, 171
in enzyme reactions 22, 23, 46, 103
of isomerizations 192, 193, 197, 198
lignan synthesis 192, 197
on liquid crystals 169
cis-methyl formation 81
and micelles 159–173
in morphine biosynthesis 29
in multistep enzyme synthesis 46
nucleophilic attack 87, 96, 98, 99, 200, 201
protein folding 181, 182
reaction trajectories 87, 93, 96, 97
for ring closure 87, 88, 91
ring opening 93

stereochemistry *(continued)*
  in Robinson ring annelation 69, 70, 71
  in sesquiterpene synthesis 71–75
  skew in biphenyl derivatives 193, 194

steroid biosynthesis
  21, 22, 82
  methyl migration 68
  Sir Robert and 68, 69, 82
  squalene 68, 82, 147
steroid synthesis
  68, 69, 73, 81
strychnine
  82, 83
sulphenamides
  62, 63

terpenes
  biosynthesis 7–9, 10, 11, 21, 22, 67
  carene 21
  cyclization 6
  isoprene rule 67
  isoprene-Ni complexes 200
  oxygenated 45
tetracycline
  53, 54, 65
theoretic organic chemistry
  107, 130
thermodynamic quantities
  MINDO/3 calculation 109–112, 119, 123
transition states
  alkyl halide 114
  calculation of 109
  on carboxypeptidase A 182
  Cope rearrangement 119, 120, 126
  Diels–Alder reaction 115, 116
  dioxetan dissociation 122, 123
  effect of solvent 182
  enantiomeric recognition 159, 160
  and micellar catalysis 153, 159, 160
  MINDO/3 109, 114, 115, 116, 117, 118
  pentacovalent 114
  ring closure 87, 98, 99
  S_N2 reactions 23, 87
tricarbonyliron
  –diene complexes 194–198
  removal 197
  stereospecific synthesis 194, 195
tropinone
  131, 132

uroporphyrinogens
  32, 34–39, 42, 43

vetivone
  71, 74, 75, 81

zinc
  in carboxypeptidase A 176, 177, 184, 186, 188